£5

Beautifully supplied
with original
engravings.

———————

The
Great Chain
of Life

by Joseph Wood Krutch

WITH ILLUSTRATIONS BY

Paul Landacre

EYRE & SPOTTISWOODE
LONDON

1957

The selection from *The Mountains of California* by John Muir, copyright 1911 by Century Company, is reprinted by permission of the publishers, Appleton-Century-Crofts, Inc.; the quotations from *Lives of Game Animals* by Ernest Thompson Seton and Roger Tory Peterson's Introduction to *Birds as Individuals* by Len Howard are reprinted by permission of the publishers, Doubleday and Company, Inc.; the quotation from *The World Grows Round My Door* by David Fairchild is reprinted by permission of the publishers, Charles Scribner's Sons. Some paragraphs in Chapter IX appeared in slightly different form in *The American Scholar*. I thank the editors for permission to use them here.

Printed in Great Britain by
Lowe and Brydone (Printers) Limited, London, N.W.10

For

WANDA AND EMERY NEFF

old friends now part of a different biota.

Prologue

WHENEVER men stop *doing things* long enough to *think about them,* they always ask themselves the question: "What am I?" And since that is the hardest of all questions to answer they usually settle for what looks easier — "If I don't know what I am, then can I tell what I am like?"

To that there are three common answers: "Like a god," "Like an animal," and "Like a machine." Perhaps there is some truth in all but the most evidently true is the second.

Man does not know how much he is like a god because he does not know what a god is like. He is not as much like a machine as he nowadays tries to persuade himself, because a machine cannot do many of the things he considers of supreme importance. It cannot be conscious; it cannot like, dislike, or desire. And it cannot reproduce its kind.

But man is so much like an animal — which can do all these things — that even the most convinced proponents of the other two answers always admit that he is something like an animal too.

Primitive man acknowledges the likeness by adopting an animal "totem," and by inventing legends which recall a time when the community was closer and more openly manifest. The most otherworldly of theologians regretfully admit that the most precious part of man, his non-animal soul, inhabits an animal body. The naturalistic nineteenth century concluded that it was from an animal that man himself had "descended."

That he did in fact so descend (or ascend) is almost certainly true whether or not this is the whole truth about him and whether or not the inferences commonly drawn are correct or adequate. If we are going to accept also the now usual assumption that man is *nothing but* an animal, then we ought to be sure that we know what an animal is capable of before we agree to the more cynical conclusions to be drawn from the common belief. The simplest purpose of this book is to suggest by concrete illustrations what being "like an animal" means, and it will stress what might be called the privileges rather than the limitations involved.

Those indignant Victorians to whom the zealous apostles of Darwin expounded their theory of the Descent of Man did not see the matter in this light. Their reaction was not unlike that of a man whose cherished pedigree has been challenged. Did not the Darwinian iconoclasts banish God from his family tree and put an ape in His place? Was not the inevitable consequence to suggest that man's qualities were not those of the noble forefather he had chosen for himself but rather those which might be expected from the base blood now said to flow in his veins?

Though we are inclined to make fun of Victorian fears, they were not entirely unjustified. Psychology, ethics, and philosophy really have tended more and more to interpret man almost exclusively in terms of what the Victorians called his baser or animal nature, and those whose business it is to contemplate or to judge him have been more and more inclined to exclaim "How like a beast" instead of "How like a god." So insistently have they stressed the less attractive aspects of both the human being and the animal that man is often driven either to despair of himself or to a cynical acquiescence in the infamy of nature.

But not everything about the beast is beastly and this large consolation is one of the things the present book offers its readers. To be an animal is to be capable of ingenuity and of joy; of achieving beauty and of demonstrating affection. These are surely not small things, though there is danger that we are forgetting how far from small they are. They are godlike attributes whether or not there is anything else godlike in the universe. To be alive at all, even if only as an amoeba is alive, is to be endowed with characteristics possibly unique and certainly exceptional throughout that vast expanse of space which extends for billions of light years beyond us, farther than telescopes — and much farther than thought — can reach.

In an attempt to understand them we shall draw upon both what the biological sciences have learned and, also, upon those direct experiences the sympathetic observer can have. We shall try first of all to look with unprejudiced eyes at specific examples of animal behavior. At the same time we shall also do a little more than that. We shall ask not only what the animal is doing but also how "aware" it seems reasonable to suppose that he is.

Science often objects to any such procedure. It sometimes insists that "behavior" is the only thing we can really know and that we should stop there. If we do not, so it often says, we run the risk of attributing to animals thoughts and feelings they do not have. Sometimes it says even that the safest assumption of all is that they have no thoughts or feelings of any kind, that they are almost, or absolutely, automata. But that assumption is actually no "safer" than the other. We have never entered into an animal's mind and we cannot know what it is like, or even if it exists. The risk of attributing too much is no greater than the risk of attributing too little.

It would, of course, be very reckless to assume that an animal's consciousness is exactly like ours. We do belong to a unique species and probably have both the keenest intelligence and the most vivid emotions (though not the keenest senses) in the whole animal kingdom. But if we really are animals, then the difference is hardly likely to be as great as the difference between sentience and automatism. If our consciousness "evolved" it must have evolved from something in some degree like it. If we have thoughts and feelings, it seems at least probable that something analogous exists in those from whom we are descended.

There are, I think, good reasons for believing that in the simpler animals consciousness is almost inconceivably dim. There is some reason for guessing that in some others, notably the insects, it has grown dimmer than it once was. But there is no good reason for doubting that many of the "higher" animals have some kind of very acute awareness which manifests itself in striking ways.

Probably their emotions are much more like ours than their thoughts can be. A dog certainly smells more keenly than we and perhaps he can be almost as happy or unhappy, as joyous or as fearful. But his intellectual insight is demonstrably not very keen. Perhaps we live in a world that is first of all a world of thoughts; he in a world of odors, sounds, sights, and emotions. But at least his world, like ours, is a world of consciousness.

This book makes no pretense at being a treatise. I am not a trained scientist; only what is sometimes contemptuously called a "nature lover." I have drawn from books written by learned experts and also upon my observation of living creatures other than man in whom I have long delighted and with whom I have perhaps more sympathy than some of those who remain austerely scientific. If I express opinions on subjects which some will maintain a mere nature lover has no right to discuss, it is because, having read much and observed a good deal, I am sometimes forced to the conclusion that the whole truth is not always represented in certain of the orthodox attitudes. The intuitions of a lover are not always to be trusted; but neither are those of the loveless. If I have also sometimes given way to that irritation which the layman often feels in the presence of the expert, I hope it will not be assumed I have forgot an essential fact, namely that I owe to the experts the technical information I appropriate.

In selecting examples of animal behavior for presentation in what is first of all a descriptive book, I have found myself usually choosing those which suggest a thesis that I hope will gradually emerge. Certain questions have been nearly always at the back if not in the foreground of my mind: To what extent is the animal that is doing any one of the thousands of remarkable things animals do aware of what he is doing? Does he always do best what seems to be consciously purposeful? And — since the answer to this second question is no — then what is the function of consciousness and why did it perfect itself in a world where, so we have been told, nothing persists except in so far as it has survival value?

We shall begin with the simplest creatures we know anything about and with the fact that they are not really simple at all. We shall then pass to others "higher" in the scale but from certain points of view hardly more remarkable. And we shall raise questions as we go along. If we end with some that are very fundamental perhaps some readers will agree they are at least legitimate questions.

Contents

The
Great Chain
of Life

1. Basic Forms of Life

EVERY SCHOOLBOY knows — or at least has been told — that our ignorant ancestors believed in "spontaneous generation." They assumed of course that all the nobler animals, including man, had to have a mother and, usually, a father as well. But the humbler creatures were so little different from the mud and slime amidst which they lived that they were assumed to be merely generated by corruption. Hence, so it seemed, some organism was taking every day the great step from lifelessness to life.

Food kept too long turned into maggots. Thousands of creepers and crawlers were generated every year when the waters of the Nile receded. "Out of strength cometh forth sweetness," said Samson — because he had seen the carcass of a lion that had apparently rotted itself into a colony of bees.

The assumption died hard. Late in the eighteenth century common sense was still defending it hotly, and one need not be surprised. Life is very persistent and very ingenious in seizing every opportunity. It takes both a sharp eye and rigidly controlled experiments to prove that "everything living comes from an egg" is a rule without exceptions. It is not too difficult to demonstrate that there will be no maggots in food protected from flies. But even humbler forms of life are harder to trace.

Put half a handful of dried grass in a glass of water, let it stand a few days until a scum has formed on the surface, and then look at a drop with even a toy microscope. Scores of little creatures will be dashing madly here and there. Several different sorts are immediately recognizable and the population will change from day to day with startling rapidity. Of course you can take a thimbleful of pond water instead, but the chances are that it will not be as teeming as the hay infusion. Moreover the quick appearance of life where none could be found before is the most startling aspect of the whole phenomenon.

Anyone who naïvely performed the experiment could readily be pardoned if he supposed that he had witnessed the creation of life itself. The dead grass decays and there is life again, generated out of corruption. But of course nothing of the sort has really happened. Sterilize the grass and the water; put them into a flask instead of a glass; insert a plug of cotton to keep airborne particles out; and nothing at all will happen.

What *did* happen was that dried though still dimly living bacteria revived in water and multiplied prodigiously, though most of them are too small and too transparent to be seen with a toy microscope. Simultaneously, a few motes in which some kind of animal life — specific for each different kind of tiny creature — had been similarly suspended sprang into life. Because some of them were animal, not vegetable, they devoured the bacteria, grew, and multiplied. Presently even carnivorous animalcules which live off those which preceded them also appeared.

But life was revived, not created. The most dogmatic adherent of the theory that once, long ago, some simple organism actually was created by some merely chemical process will have to admit that, so far as he knows, the miracle never happens now, and that indeed all evidence points to the conclusion that probably, during all of geological time, it never happened but once. It was, as they somewhat lamely confess, a very "improbable" chemical reaction.

By the time Charles Darwin was born, "spontaneous generation" had ceased to be a respectable theory. It was generally admitted that life had begun a long time ago and that its origin was a tremendous event, not something which was repeating itself daily in every body of stagnant water or every piece of decaying meat. By his time interest was shifting from *origins* to *development*. And since it was very convenient to have something very simple to begin with there was nothing more inviting than the assumption that the tiny specks of living material, then commonly called "infusoria" because they appeared in an infusion of hay, were surviving examples of what nature's first experiment with life was like. They seemed barely living and no more. The bridge between them and the merely chemical ought not be too difficult for the imagination to bridge and once you had got across it the story would be complete. Evolution could tell it "from Amoeba to Man." Presumably life started with what came to be called the protozoa, or pre-animals, and ended, at least for the present, with the human being.

No wonder, then, that some incautious early writers talked about the simplicity of these microscopic creatures, that they sometimes spoke of them as though they were mere blobs of undifferentiated protoplasm, and liked to think them essentially similar to the cells that compose the bodies of higher animals.

Actually and even before Darwinism intensified the search for a missing link between animate and inanimate it was beginning to be realized that the animalculae were not so very simple after all. And the more these animalculae are studied, the more evident it becomes that the concept of un-differentiated protoplasm is largely fanciful. Unlike us and the other higher animals, they have no lungs, hearts, muscles, glands, or other elaborate, cell-built machines. But that does not mean that only elementary life processes can be carried on. Even some of the simplest have, like the classic amoeba, an elaborate physiology. Many others exhibit complicated structures — tough skins, mobile hairlike cilia, sometimes even hard skeletons — all of which are the more remarkable because they are not constructed out of blocklike cells but are simply part of the by no means undifferentiated blob of protoplasm. The very word "protozoa" (coined in 1817 and meaning "pre-animal") is misleading because they are not "pre-animals" but in every significant respect merely "ani-mals" instead. And there are at least fifteen or twenty thousand kinds, differing so widely in structure, physiology, and habits of life that most generalizations about them are subject to exceptions. They differ among themselves far more than fish differ from other fish, reptiles from other reptiles, or mammals from other mammals.

Without stomachs they digest; without lungs they take in oxygen; without kidneys they secret uric acid; and without bladders they collect it in pockets from which it is finally expelled. Without sex organs they have an effective system for mingling heredities, and they exchange portions of chromosomes carrying hereditary traits quite as effectively as the most highly developed animals. They are tiny but just as difficult to explain as a man or a whale. Simple, indeed! If the first living things really were like them, then the sudden appearance of a protozoan was a phenomenon almost as astonishing as the sudden appearance of an elephant would have been.

Before these facts were known it was natural to associate the more complicated physiological processes carried on within the higher animals with the mechanisms that perform them. But as the protozoan demonstrates, breathing is possible without lungs, movement without muscles. The ability to live and to perform the actions associated with living is not dependent upon the known mechanisms which thus seem to be revealing themselves as incidental, not essential.

As a matter of fact the animalcule which lives without them is even harder to understand than the animal which has them at its disposal. Many cells in a human body seem mere building blocks; but the single-celled animal is astonishingly complete. Unlike some immediate successors, Anthony van Leeuwenhoek, the seventeenth-century maker of his own very good lenses and the first to pay serious attention to these "animalculae," assumed that they must be provided with miniature replicas of the familiar animal organs. "How marvelous," he exclaimed, "must be the visceral apparatus shut up in such animalcula." The fact that they actually get along without this "visceral apparatus" is even more remarkable.

Consider, for example, the case of a common sort which, like various others, has a permanent mouth and throat — that is, a tube-like opening from the outside kept open by a cyclindrical arrangement of rigid rods that simply appear in the material of the single cell. Like so many animalcules, this creature multiplies by division — which is to say that from time to time two dents appear on either side of it somewhere near the middle and then cut gradually inward, rather as though a loop of thread were being tightened around it. In the kind we are discussing this means that of the two newly formed individuals one will have a throat complete with reinforcing rods, while the other will not. What will it do about the predicament in which it finds itself?

It so happens that this situation is one I have observed for myself and I do not think there is any wonder in nature more wonderful than what goes on within this tiny cell. The mouthless and throatless individual simply makes — one is tempted to say "wills" — another mouth and a throat. An opening begins to appear; the reinforcing rods begin to take shape. In about fifteen minutes the whole structure is indistinguishable from that which his twin received ready-made.

All growth and development are surprising enough. The fertilized cell on the surface of a hen's egg which grows, step by step, until soon there is a beating heart, is miraculous. But we seem to understand it in some mechanical sense because cell is added to cell and a machine which begins to function is created. The very existence of that machine distracts our attention from the fact that it built itself.

But in the case of my animalcule there is no such distraction. Its throat and mouth were not built up from visible blocks. There is no mechanism which can be understood. It seems almost as though something, either the animalcule itself or something outside it, had said, "Hocus-pocus: let there be a mouth." And in a few moments a mouth was there.

In all the higher animals certain cells are specialized for the sole function of muscular contraction or for the transmission of nerve impulses. But in the one-celled animal the organism functions as a whole. And what this fact suggests is simply that a muscular system and a nervous system are conveniences, not necessities. We do not have the ability to react *because* we have a nervous system but, on the contrary, have a nervous system because protoplasm itself is capable of reaction.

For a similar reason it is possible to feel that the optical instrument we call an eye, and the *mechanism* which can be explained in terms of a lens that forms an image exactly as the lens of a camera does, is less mysterious than an animalcule which "sees" — at least to the extent of being aware of light and of moving toward it — without having an eye at all. Some of the very "simplest" or "lowest" carry a little spot of color which is structurally not an eye at all, merely a light-sensitive area. But the tiny creature can be "blinded" if the spot is destroyed and will no longer move toward the light.

Every capacity exhibited by a complicated body-machine seems potential in the mechanically simple protozoan cell. It is easier to think of a man as essentially a mechanism than it is to think the same of a protozoan. In it life itself seems less dependent on mechanical contrivances than in any "higher" creature.

There is an old story about a nineteenth-century physicist who complained that the human eye was a very clumsy optical instrument and that he could make a better one. The classic reply, "But you could not make it see," gains new point from what is known about the protozoan "eye." Animals, so it seems, did not learn to see because they developed what we call an eye. On the contrary, they developed eyes because they were already able, in some sense, to see. And this ability to see is not in any way explained or accounted for by the fullest understanding of the optical principles involved in the highly developed eye itself.

Over and over again modern man has reinvented the mechanical contrivances that nature had hit upon millions of years before his time. He put valves into his water pumps which work on the same principle as those in the mammalian heart. He discovered the mechanical uses of the lever several geological epochs late. Very recently he learned to make artificial silk by extruding from a fine orifice a substance that hardens immediately on contact with the air —

which is precisely what spiders had been doing as far back, perhaps, as the Paleozoic.

But no example of this johnny-come-lately inventiveness is more striking than the photographic camera, which as a mechanical contrivance imitates so precisely the animal eye — in everything, that is, except the ability itself to "see." The fact that in the higher animals nature's eye is not only a camera but a "three-color camera" and that all our recently invented processes for color photography imitate its method by providing for a separate image of each of the three primary colors and then leave it up to the brain to "see" them, not as three primary colors but, by combination, as the whole spectrum, only reinforces the lesson. We can make the same machine that nature made. But we can't make our cameras *see* any more than we can make our telephone systems talk to themselves, or our electronic calculators propose any problems without the intervention of something we cannot invent.

Nearly fifty years ago the Cambridge protozoologist Clifford Dobell protested against the manifest absurdity of pretending that the microscopic animals were fundamentally simple, that they could be taken to represent what life was like at the dawn of creation, or, even, that there was any sound reason for calling them "lower." He objected to calling them "one-celled" and proposed "noncellular" instead. "Smaller," he added, was the only adjective that could legitimately be employed to distinguish them, and though this is perhaps only a corrective exaggeration, no responsible biologist would today accept "blobs of undifferentiated protoplasm" as an admissible description — though many laymen probably still tend to assume what would greatly simplify things if we could assume, namely, that "the simple one-celled animals" represent life at a level so low that they are something like what it was at its first emergence.

What any real study of these "small" animals does suggest is, on the contrary, that they probably represent the end result of a long evolutionary process and that they are so far from being just barely alive that many of the characteristics which distinguish living from nonliving matter are as unmistakably present as they are in the human being.

Thus the simplicity of even the amoeba is mechanical, not vital. It does not come equipped, as the higher animals do, with a complicated pump for a heart, an elaborate camera-like eye, or an elaborate communication system like the brain. But it is already endowed with characteristics that make it as different from the most complicated machine as man himself is different from it. And because of this fact it is much easier to imagine that an amoeba might in the course of time develop an eye and a brain than it is to imagine that a camera could learn to see or an electronic calculating machine could ever, in the course of no matter how many millions of years, develop consciousness or a will-to-live. As the Johns Hopkins biologist Jennings — certainly not a mystical writer or a man given to fantasy — once wrote: "If Amoeba were a large animal, so as to come within the everyday experience of human beings, its behavior would at once call forth the attribution to it of states of pleasure and pain, of hunger, and the like, on precisely the same basis as we attribute these things to a dog."

As a species, amoeba is obviously much older than either man or any other "higher animal." The latest estimates of the earth's age allow for a lapse of a billion years or more between the time when the first rocks were formed and the time when the first fossils were laid down. Hence if there ever was any such thing as that highly hypothetical "undifferentiated blob of protoplasm" then it may well have been given a far longer stretch of time in which to become an amoeba than any amoeba-like creature was given to become a man — and the achievement may well have been even more impressive. A great deal of ink has been spilled over the "missing link" between the anthropoids and *Homo sapiens*. But it is as nothing by comparison with all the links which are missing — if they ever existed — between amoeba and that first particle of barely living matter which, as so many glibly assume, arose as the result of an admittedly "improbable" (and never repeated) chemical reaction.

Since the discovery of the viruses — presumed organisms too small to be seen with an optical microscope — there has been an inevitable tendency to suggest that perhaps *they* represent the very earliest form of life. If the protozoan is not a simple blob of protoplasm, perhaps the virus is the missing link between the living and the nonliving.

But inevitable as this guess is it is not more than a guess and objections can be raised. Viruses cannot reproduce except by living in more highly developed organisms, and the very first form of life can hardly have been one dependent upon a more highly developed form! Moreover, this dependence upon a host suggests parasitism and it is generally believed that parasites are not primitive but are degenerate plants and animals which devolved into simplicity after they fell into the reprehensible habit of living off others. All this seems to make the guess that viruses are degenerate rather than ultraprimitive forms of life at least as persuasive as any other.

Try to begin the story of life and its various adventures at the real beginning and you are thrown back upon pure speculation. The origin of life is as much a mystery as it ever was. In fact it seems even more mysterious than it did in the days when spontaneous generation was supposed to be an everyday occurrence; something which did happen rather than something which happened once but could not happen now. Consent, on the other hand, to begin not at the beginning but with life in the simplest known form as a *fait accompli;* accept as your starting point not a chemical reaction, but a small creature already endowed with the essential characteristics of all living creatures even though his body machine is relatively very simple, and it is possible even at this late date to trace what seem to be some of the steps by which the mechanism was elaborated and the potentialities inherent in life itself released as it provided itself with the body mechanisms it had by then learned to manipulate.

Grant the ability to breathe, and the stages by which lungs came into being can be demonstrated in a probable-seeming sequence. Grant the ability to "see" in even the elementary sense that a protozoan moving toward an illuminated area may be said to see, and the stages by which an eye seems to have been developed can be demonstrated in the same way. That makes a very interesting tale and one far more convincing than any which can be told by those who insist upon going back to a purely hypothetical "origin of life" instead of beginning in the middle of things — which is precisely where we find the one-celled animals.

Pond water and a toy microscope — though I would certainly recommend a better one — will reveal some of these stages as clearly as they revealed the fact that in the simple protozoan it is only the apparatus which life there employs, not the life processes themselves, that is simple at all. The first thing the microscope will reveal is that "one-celled" and "many-celled" are not an absolute distinction and this fact will carry us across one of the most difficult of all the bridges between a simple body-machine and a complicated one.

Something not unlike an amoeba flows with the blood stream in every human being. Like an amoeba it is a speck of protoplasm not confined within a rigid cell wall. It seems to live a semi-independent life and, fortunately for us, it can, again like an amoeba, engulf and then digest bacteria. When we find it inhabiting a mammalian blood stream we call it a "phagocyte" or white blood corpuscle.

But most of the cells composing the body of any higher animal are much less like an individual protozoan. It is true that they consist of protoplasm surrounded by a retaining wall like that of many one-celled animals. It is also true that each possesses a mysterious nucleus with its even more mysterious chromosomes, and that this nucleus seems to be the very inner sanctum of life. But most of these cells are incapable of leading even a semi-independent life. They cannot, like a protozoan, perform a whole repertory of functions. Many can do only one thing — say contract as part of a muscle or transmit a nerve impulse as part of a nervous system, both of which remarkable things a single protozoan body can also do.

The difference is so great that these specialized cells seem in one sense degenerate, and by some it has been maintained that the whole theory that represents them as modified one-cell organisms is insupportable. But if the theory actually is correct, then the fact that there are in pond water certain creatures not clearly either one-celled or many-celled does suggest the stages by which the situation existing in the body of all the higher animals may have come about, though the revelation of these stages leaves the question "how" as puzzling as ever.

Spend a little time with your pond water, spend enough to begin to recognize some of the innumerable forms of life which may turn up in it. You will soon notice that certain one-celled animals occur always as single independent individuals and that when they multiply by division the two into which one seems so easily to have turned itself immediately separate to go their individual ways. On the other hand there are others which tend to cling together and form colonies — sometimes in the form of a chain, sometimes as irregular aggregations.

So far as is known, the contact is purely physical. Each individual appears to be perfectly capable of living an independent life. One does nothing the other does not do. They are merely, in some haphazard way, gregarious, though there is no division of labor. Other species form more definite colonies, sometimes embedded in a sort of jelly which gives a semipermanent identity to the community itself. And though it is usually assumed that each individual cell is still capable of independent life, there may possibly be a tendency on the part of some members of the group to specialize, at least temporarily, in certain functions.

Many of these unenterprising colonies are rather dull to watch because they don't seem to do much except vegetate unambitiously. But if you are lucky, in your pond water you may chance upon a superficially striking and, if closely observed, one of the most crucial of all the colonial forms of small plant-animals — crucial because it exhibits a startling advance in what I have chosen to call the elaboration of a body-machine.

This astonishing organism — named long ago *Volvox*, or the Roller — is visible to the naked eye as a tiny green speck. Magnified only a hundred diameters or less it reveals itself as a spherical green jewel. If you have been careful to allow for its comparatively large size by suspending it in a drop of water hanging from the bottom of a very thin piece of glass, it will justify its name by rolling majestically.

Slow its motion by introducing into the drop of water a trace of some gummy substance like gum arabic or even gelatin and it will be evident that it is composed of a large aggregation of cells not all of which are alike. Each of those which occupies a place on the periphery of the sphere is provided with a pair of hairlike flagella, or whips, precisely like those many one-celled organisms possess, and the thrashing of these whips is responsible for both the revolving and the forward motion of the whole group. Within the sphere most of the cells have no such appendages, but they are not all alike. Here and there are even more mysterious groups of what appear to be still more highly specialized cells of several different kinds. If Volvox is a colony then the individual organisms which compose it have obviously gone in for a rather high degree of specialization and co-operation.

Volvox is, in other words, rather more a many-celled individual than a colony of separate one-celled organisms. And when it became such it took a long step toward mechanical elaboration. The most obvious evidence of the fact that it certainly did take this step is the co-ordination that exists among the hundreds of whips which drive it on its course. They do not wave at random. They are synchronized in such a way that they seem to be animated by a single purpose and to be directed from a central seat of the group will. Threads of protoplasm, which are not nerves but obviously function as such, connect one whip-bearing cell with another, and along these threads an impulse must travel. Volvox has no brain but it does some kind of thinking without one.

To most biologists this handsome little creature is taken to be the simplest now-surviving organism of which it can be said that it is definitely multicelled and not a mere association of one-celled individuals. Since it has no limy or bony shell or skeleton neither Volvox nor any similar creature has ever left a fossil behind. How different it may be from some predecessor no one can profitably guess. But the simplest of the less simple organisms that did leave fossils and are also among those still living today seem to have changed comparatively little, and it is not a wild guess to suppose that something a good deal like Volvox was already rolling purposefully through stagnant waters many hundreds of millions of years ago — already thinking without a brain and probably doing it a little better than any one-celled organism could.

This, so biologists hasten to assure us, does not mean that man or, for that matter, the fish and the reptiles are directly descended from Volvox. As one of them has put it, Volvox represents a "phylogenic impasse," which translated means, "an experiment that never got any farther." Yet most are willing to assume that the higher animals did begin as something somewhat similar to the protozoa of the present day.

If all of this is true then Volvox, as presumably the simplest now-living creature which took the crucial step in the direction of that kind of body-machine characteristic of what we call the higher animals, is worthy of great fame and considerable attention.

For purely mechanical reasons no one-celled animal can be very large. Had something not hit upon the device Volvox exemplifies, no individual living creature could be much bigger than a few hundredths of an inch long at most, and probably it could never have made any considerable technological advance beyond that which the body-machine of the more complicated protozoa now exhibits. In addition to this Volvox has also two other and even more startling "firsts" to its credit.

No simpler creature develops special cells distinguishable as male and female; no simpler creature ever dies of natural necessity.

Surely the introduction into our universe of Sex on the one hand and of Death on the other is worth a new chapter.

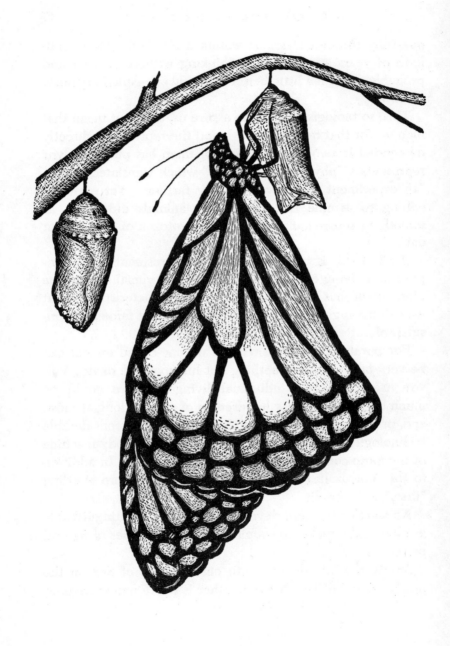

2. Machinery for Evolution

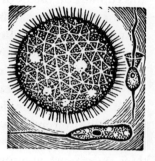

ON THE SECOND of January 1700 Anthony van Leeuwenhoek, draper of Delft and self-taught Columbus of the littlest world, was writing to the Royal Society of London one of the many letters in which he described his voyages of discovery within a drop of water.

To William Dampier and other such rovers he left the exploration of the terrestrial globe. To another contemporary he left those equally adventurous voyages "through strange seas of thought, alone" which took Newton across abysses of space to spheres much larger but not so little known as those to which Leeuwenhoek devoted his long life.

These worlds of his were not lifeless but teaming with life; and his discoveries, unlike those of Columbus, were discoveries in an absolute sense. He saw what no man, not merely no European man, had ever seen before. He had every right to be — he probably was — more amazed than Balboa.

The draper of Delft was already just short of seventy when he wrote:

> I had got the aforesaid water taken out of the ditches and runnels on the 30th of August: and on coming home, while I was busy looking at the multifarious very little animalcules a-swimming in this water, I saw floating in it, and seeming to move of themselves, a great many green round particles, of the bigness of sand-grains.
>
> When I brought these little bodies before the microscope [actually a single very small lens which he had ground himself and fixed between two perforated metal plates] I saw that they were not simply round, but that their outermost membrane was everywhere beset with many little projecting particles, which seemed to me to be triangular, with the end tapering to a point: and it looked to me as if, in the whole circumference of that little ball, eight such particles were set, all orderly arranged and at equal distances from one another: so that upon so small a body there did stand a full two thousand of the said projecting particles.
>
> This was for me a pleasant sight, because the little bodies aforesaid, how oft soever I looked upon them, never lay still; and because too their progression was brought about by a rolling motion. . . .
>
> Each of these little bodies had enclosed within it 5, 6, 7, nay some even 12, very little round globules, in structure like to the body itself wherein they were contained.

There is no mistaking the fact that what had just swum into Leeuwenhoek's ken was that very original and inventive organism, Volvox. Through hundreds of millions of years it had waited in countless places for man to become aware of its existence and, ultimately, to guess how important a step it had taken in the direction of both consciousness and that curiosity which was leading the Dutch draper to seek out in the ditches of Delft. The "little projecting particles" are the peripheral cells which enclose the watery jelly in Volvox's interior. The "5, 6, 7, nay some even 12, very little round globules, in structure like to the body itself" are the vegetative "daughter cells" produced by a sort of virgin birth between the sexual generations, much as in some of the higher plants "offsets" as well as seeds are produced. Nor did Leeuwenhoek's observation stop there:

> While I was keeping watch, for a good time on one of the biggest round bodies . . . I noticed that in its outermost part an opening appeared, out of which one of the enclosed round globules, having a fine green color, dropped out; and so one after another till they were all out, and each took on the same motion in the water as the body out of which it came. Afterwards, the first round body remained lying without any motion: and soon after a second globule, and presently a third, dropped out of it; and so one after another till they were all out and each took its proper motion.
>
> After the lapse of several days, the first round body became as it were, again mingled with the water; for I could perceive no sign of it.

In other words Leeuwenhoek saw both the liberation of the daughter colonies and also, though he did not realize its importance, something even more remarkable. He saw Volvox yielding to one of its two remarkable inventions — natural (or inevitable) Death. The other half of the strange story eluded him completely. He did not know that after a few generations have been vegetatively reproduced by the process he observed there comes a generation that will produce eggs which must be fertilized by sperm before they can develop.

Nearly three hundred years later and more than five thousand miles from Delft, I, in my late turn, have also been looking at Volvox still rolling along in his gracefully expert way. Like most of the free-living protozoa, he has established himself pretty well over the whole earth outside the regions of eternal ice and there is no use speculating how the cosmopolitan distribution was achieved or how he got to America. During the vast stretches of time which have been his, routes were open at one time or another from every part of the earth to every other part.

Historical plant geographers come almost to blows over the question how, for instance, the sweet potato got to the South Sea islands. But Volvox's history goes back too far for even speculation or contention to reach. If only the fit survive and if the fitter they are the longer they survive, then Volvox must have demonstrated its superb fitness more conclusively than any higher animal ever has.

My equipment is as much superior to Leeuwenhoek's as his originality, ingenuity, and persistence were superior to mine. Instead of a single blob of glass fixed in front of a tiny tube holding water and held up to sun or candlelight, I use the compound microscope, which was not brought to its present state until the second half of the nineteenth century. Light passes from a mirror through a complicated set of lenses designed to place it at exactly the right spot. An image is formed by another series of lenses, cunningly designed to correct one another's faults and then form an image in a black tube, this image again magnified by another set of lenses. I can confine Volvox to a hanging drop of water; I can light him from below, from the side, or even from the top. I can slow him down with sticky substances introduced into the drop; and I keep a supply of his species thriving in an artificial culture medium prepared for me by a biological supply house dealing in all sorts of improbable things. But though the beauty of Volvox must be even clearer to me than it was to Leeuwenhoek I have seen only what he saw and described in unmistakable terms.

Under a magnification of no more than a hundred diameters — called by microscopists "very low power" — Volvox looks about the size of a marble and when motionless less like a plant-animal than like some sort of jeweler's work intended, perhaps, as an earring. The surface of the crystal sphere is set with hundreds of tiny emeralds; its interior contains five or six larger emeralds disposed with careless effectiveness.

But Volvox is seldom motionless when alive and in good health. Bright as a jewel, intricate as a watch, and mobile as a butterfly, his revolutions bring one emerald after another into a position where they sparkle in the light. Though he is called the Roller, he actually *revolves* rather than *rolls*, because he seems to turn on an invisible axis, much as the planets do, and he moves forward with this axis pointing in the direction of his motion. His speed varies and he frequently changes direction but I once counted the seconds it took him to cross the field of the microscope and calculated that his speed, in proportion to his size, is comparable to that of a man moving at a fast trot. Volvox, however, suggests nothing so undignified as a trot. There is something majestic and, one might almost imagine, irresistible about his revolutions — again like those of a planet. One half expects to hear some music of the spheres.

Because the microscope has temporarily abolished the barrier of size which separates the universe of Volvox from my own, I enter temporarily into a dreamlike relationship with him, though he is unaware of my world and perhaps equally unaware of his own. Nevertheless there is an easy purposefulness in his movements and from what I have learned from the careful research of others I know that it is all much more complex and astonishing than I would ever have guessed or than Leeuwenhoek did guess.

When I lift my head from the microscope the dream vanishes. But it is man and his consciousness which is really the fleeting dream. Volvox, or something very much like him, was leading his surprisingly complex life millions of years before man's dream began and may well continue to do so for millions of years after the dream ends. Harder to realize is the fact that the enterprise and adventures of Volvox typify certain of the innovations and inventions which are casually summed up in the word "evolution" and hence constituted some of the earliest and most essential steps toward making possible our dream.

Such acquaintance with Volvox as I have gained from my

own casual observations is as superficial as what one gets from a moment's examination of a flower or from peering with mild curiosity at some strange animal behind the bars of a zoo. Yet even so casual an examination will lead one to guess at some of the significant facts.

The predominant color of Volvox is green; the green looks like chlorophyl, and so it is. Moreover Volvox can be "cultured" in a purely chemical solution. All of this suggests a plant rather than an animal, but biologists have decided that the classification is meaningless at this level. Both textbooks of zoology and textbooks of botany usually claim Volvox, and there are no hard feelings because zoologists and botanists agree that Volvox represents a stage of evolution at which plants have not diverged from animals. He is either the one or the other. Or, more properly, he is neither. He is not a plant or an animal; he is simply something which is alive.

Concerning another ambiguity the observer is likely to make an even better guess. The walls separating one cell from another are clearly marked out and each of the cells which lie upon the periphery of the sphere is exactly like the other peripheral cells. Every one of them looks like a complete one-celled creature, lashing its two flagella precisely as many such one-celled creatures do — although, as was mentioned a few pages back, it is easy to see that Volvox's cells do not crack their whips independently because they are co-ordinated in such a way that the organism as a whole moves purposefully forward in a given direction as though the individual flagella were controlled by a central intelligence.

Yet the individual cells not only look like separate animals but are in fact morphologically all but indistinguishable from a certain very common specific free-living organism that occurs by the millions in stagnant rain water and is one of the usual causes why such water takes on a green color. Like this creature each of Volvox's peripheral cells has a nucleus, two flagella, and an "eye spot." Any observer might easily suppose that a number of such tiny creatures had recently got together and formed a ball; or, that the one-celled creatures were the result of the dissolution of Volvoxes. But if the one-celled creatures did form aggregates which then became cooperative, all that happened millions of years ago and for a very long time Volvox has been much more than a mere casual grouping. One of his peripheral cells cannot live without the others. He is at least as much one creature as many and he must live or die as a whole. He can no longer dissolve into the myriad of individuals of which, long ago, his ancestors were no doubt composed.

If you want to call him a multicelled animal you have to admit that the peripheral cells have retained a good deal of their original equipment for independent life. On the other hand if you want to call him a mere aggregation you have to concede that the individuals composing the colony have been co-ordinated, disciplined, and socialized to a degree no human dictator has yet even hoped to achieve in his "monolithic" state. But in any event the guess that Volvox represents a stage in the development of the "higher" multicellular animals and suggests how the transition was made is more immediately persuasive than many of the other guesses biology finds itself compelled to make.

Without comment I pass over the suggestion made sometimes with horror and sometimes with approval that our present-day society is in the process of taking a step analogous to that once taken by Volvox; that just as the one-celled animal cooperated until he was no longer an individual but part of a multicelled body, so perhaps the highest of the multicelled animals is now in the process of uniting to make a society in which he will count for as much and as little as an individual cell counts for in the human body.

Now comes the most powerful argument of all for calling Volvox a unified individual rather than even a tight social group and it has to do with three different sorts of cells sometimes found within his central jelly. The least remarkable of the groups of special cells are those composing the "daughter colonies" which Leeuwenhoek saw and which in time will break out of the parent colony to start life on their own. About them there is nothing so very surprising, since "budding" of one kind or another is not uncommon among microscopic organisms. The other two groups of specialized cells are much more interesting because they seem to represent the first appearance of sexual differentiation.

One group is of spindle-shaped cells very much like the sperm of the higher animals. The other is a sort of egg considerably larger than the sperm because, like most eggs, it contains a rich store of reserve food to nourish a growing "embryo." Sooner or later a sperm cell will seek out an egg, the two will fuse, and thus they will pool the hereditary characteristics carried by the sperm and by the egg. Thus Volvox introduces a method of procedure almost universal among the higher plants and animals. Moreover, there is even the beginning of the distinction between Male and

Female *individuals* as well as a distinction in the sex cells themselves, because, in the species which I have been observing, a given individual usually produces either eggs only or sperm only. The cynic who said that the two great errors in creation were the inclination of the earth's axis and the differentiation of the sexes probably had no idea how long ago the second error was made. To eliminate it we would not have to wipe the slate quite clean but we would have to go back at least as far as Volvox.

Being the inventor of sex would seem to be a sufficient distinction for a creature just barely large enough to be seen by the naked eye. But as we have already said, Volvox brought Natural Death as well as Sex into the world. The amoeba and the paramecium are potentially immortal. From time to time each divides itself into two, but in the course of this sort of reproduction no new individual is ever produced — only fragments of the original individuals, whose life has thus been continuous back to the time when life itself was first created. Though individuals can be killed there is no apparent reason why amoeba should ever die.

Individuals have been kept alive in laboratories for years by carefully isolating one-half of the organism after each division. What memories an amoeba would have, if it had any memories at all! How fascinating would be its firsthand account of what things were like in Protozoic or Paleozoic times! But for Volvox, death seems to be as inevitable as it is in a mouse or in a man. Volvox must die as Leeuwenhoek saw it die because it has had children and is no longer needed. When its time comes it drops quietly to the bottom and joins its ancestors. As Hegner, the Johns Hopkins zoologist, once wrote: "This is the first advent of inevitable nat-

ural death in the animal kingdom and all for the sake of sex."
And as he asked: "Is it worth it?"

Nature's answer during all the years which have inter-
vened between the first Volvox and quite recent times has
been a pretty steady *Yes*. Sex plays an essential part in the
vast majority of all the forms of life presumably more recent
than Volvox. In your very back yard, if you have one, there
is on the other hand evidence that she may be reconsidering
her momentous decision — though for the moment we shall
say no more about this startling fact.

Suppose we look instead a little more deeply into the
meaning and purpose of that sexual love for the sake of
which Volvox consented to die.

No small part of all fiction and poetry is concerned with
the various ways in which a member of one sex has "proposed" to a member of the other — with the terms in which
he (and sometimes she) has made the proposal, with the
reasons given why the proposal should be accepted, and
with the consequences that were supposed to follow if it was.

Arranged in an ascending series, such typical "proposals"
have been: "Let us share a delight," "Let us commit ourselves to one another for ever," and "Let us unite our souls."
Occasionally it has even been "Let us create a new life since
we ourselves must die." But never, I imagine, has it ever been
"Let us exchange our chromosomes." Yet that, if you insist
upon equating everything with its origins, is what sex is
really about.

These chromosomes are something with which all the
higher plants and animals, as well as the majority of all the
protozoa, come equipped. One of the reasons why the latter

are not mere "blobs of undifferentiated protoplasm" is that embedded in the protoplasm is a little nodule of material quite different from the rest of the protoplasmic mass. This nucleus, as it is called, is so tremendously important that the very secret of life seems to reside within it. A somewhat similar nucleus is found within every living-tissue cell of the human body and our nuclei are like those of many protozoa in another important respect. Both contain within themselves a number of little threads called chromosomes which seem to be the nucleus' Holy of Holies just as the nucleus is the cell's. The number of chromosomes a protozoan has is not the same as the number we have. In fact every plant and animal has a number specific for that organism and every single cell (except the sex cells) composing the tissue of any multicellular organism has the number characteristic of it.

When a typical protozoan decides that it would rather be two protozoa instead of one — when, in other words, it reproduces by fission — the first thing it does is precisely what a cell in your body does when any one of your tissues grows. No new cell would be any good without a nucleus or without a nucleus provided with chromosomes. What happens in order that it should have one is extremely complicated, but it may be simplified thus: before the cell itself begins to divide the chromosomes move about until they are arranged in a sort of spindle-shaped group. Each chromosome, already split lengthwise, cuts itself in two near the middle; two groups are formed; when the whole cell divides, one-half of each chromosome is found within the two cells existing where only one existed before.

Thus it is that the protozoan reproduces itself and that a bit of your tissue grows. The new protozoan or the new cell does not differ from the old one. What you have is merely a pair of identical twins.

Most of the cells in your body reproduce indefinitely in just that fashion. But most protozoa "conjugate" as well as divide. From time to time two meet, their bodies either completely fuse or, in the case of many, their mouths are pressed together for a considerable time in a sort of soul kiss during which each gives up to the other a portion of its nucleus. Presently each individual combines the gift portion with the retained portion of his original nucleus to make a new one. And when the time comes for division it is this new nucleus that will be shared between the two halves of itself. Hence the new individual will, like the offspring of one of the higher animals, "inherit" from each of its conjugationed parents; and because of the combination of inherited traits it will be a unique individual.

Should we, then, say that sex already exists in a large number of protozoa? The answer to that question is yes and no. It is yes because a mingling of heredities — which is the grand purpose of all sexual processes — is to some extent accomplished. Though there are no parents in the usual sense because the parent splits in two, the fact remains that the two individuals resulting from a division after a conjugation are both different from what they would have been if no conjunction had taken place. And for that reason such reproduction is commonly called "sexual," to distinguish it from either a division or a budding not preceded by any form of conjugation.

Yet the answer is no if you are thinking in terms of the complete sexual process as it occurs in the higher plants and animals. The most obvious incompleteness is in the absence of any differentiation of the sexes. No detectable difference in structure exists between two conjugating protozoa. There are no males and no females. Neither are there any specific sex cells — no eggs and no sperm. It is in respect to these two facts that Volvox takes a great step toward sexuality as it is commonly known.

Volvox never indulges in the kind of conjugation we have been describing. Neither does it ever divide into two halves. Instead it produces what may properly be called "children" — sometimes by the vegetative process which Leeuwenhoek described and sometimes by a more surprising method which no free-living, single-celled animal ever practices.

Somewhere inside its sphere appear certain groups of small cells which might at first sight be mistaken for vegetative buds. But they develop quite differently in either one of two ways. Sometimes they become quite large. Sometimes on the other hand they split up into a great number of extremely small mobile cells. The first are eggs; the second sperm. Neither is good for anything by itself. But each is ready to do what the egg and the sperm of all the higher animals do. The egg waits. The sperm seeks it out. Then the two fuse and the fertilized egg is endowed with the hereditary traits contributed by the sperm as well as with those originally its own. Some species of Volvox are hermaphroditic as many lower organisms are. A single individual, that is to say, produces both male and female cells. But Volvox is also inventing the sexual differentiation of the whole organism. Certain species are commonly either male or female; the one producing only eggs, the other only sperm.

In what ways, one is bound to wonder, is this differentiation of sex cells and the further differentiation of male and female organisms superior to the simpler arrangement which the protozoa have managed to get along with during millions of years?

It has been, as satirists have so frequently pointed out, the cause of a lot of trouble in the world. Yet there must be compensating advantages, because as one moves upward along the evolutionary scale sexuality becomes universal and even hermaphroditism tends to disappear.

That sexual differentiation provides a richer emotional experience is a reason that few biologists are likely to admit as relevant and indeed it would be hard to prove that Volvox finds life more colorful than a paramecium does. Hence the biologist has to fall back upon such things as the superior viability of an egg, which can be heavy with reserve food resources because it does not have to be active when a small mobile sperm is there to seek it out. Possibly another fact is even more important. In all the higher animals sperm and egg cells differ from every other cell in the body of that organism in that they have only half the normal number of chromosomes and that the normal number is re-established when the sperm's half-number is added to the egg's half-number — which arrangement certainly shuffles hereditary characteristics more thoroughly than when the offspring has the whole inheritance from both sides of the family. In any event (and to repeat) there must be some advantage, since every animal above the protozoan level tends to adopt the novel arrangement first observable in Volvox.

Having reminded myself of all this, I cannot resist the temptation to push the typewriter aside to stare again at Volvox, quietly revolving in the beam of light concentrated from below and directed upward through my microscope. He (or she, for there are no sex cells visible at the moment) revolves steadily like some planet; and at a magnification of one hundred diameters looks as large as Mars through the largest telescope. But no more than Mars will Volvox abide my question. Over and over again I catch myself referring to him as "simple." Yet though his organism as a whole may reasonably be called relatively simple his sexual processes are not simple, either comparatively or absolutely. In many respects they seem quite as involved as those in a human being, although his chromosomes are no doubt far less complex. Something unimaginably complicated and subtle goes on when his sex cells isolate the hereditary uniqueness of the one organism and then, later, unite it with the uniqueness of an individual of the opposite sex.

Is, I wonder, the difference between Volvox and a human being as great as the difference between it and any theoretical, half-living blob of undifferentiated protoplasm would be?

Suppose that some biochemist should realize his dream and synthesize in his laboratory something which is undeniably protoplasm and is alive. How long would it take that protoplasm, either with or without the chemist's aid, to get a nucleus with chromosomes and to acquire the habit of sexual reproduction? Is it as easy to imagine how the simple product of "an improbable chemical reaction" could invent sex as it is to imagine how something which had once got as far as Volvox has might go on to become a man?

Even for the sake of picturesqueness and drama I would hesitate to say that Volvox invented Love as well as Sex. What he invented was perhaps at most only the possibility of Love. But in even the simplest creature Death — either accidental as it always is in one-celled creatures or sometimes natural as it may be in Volvox — is all too plainly like death in every other creature up to man. It is something we have not been able to change or complicate any more than we have been able to abolish it. Volvox may not love as we love, but he seems to die as we die.

A drop of water suspended in the intense beam of a microscope is not a very favorable environment for any organism, and several times I have seen a Volvox die. Usually his activity slows down and then stops. There comes the moment when he gives up his little ghost, though the body, like a human corpse, may seem hardly different from what it was before. A tiny spark is extinguished as suddenly and as irrevocably as the larger spark in any larger animal. What was alive is dead. The immeasurable, indescribable difference between the animate and the inanimate has been produced in a single instant. The spark can no more be revived in Volvox than in man. He was rounder than Humpty-Dumpty but not all the king's horses or all the king's men. . . .

It is no wonder, I say to myself, that so many men in so many different places and at so many different times have assumed that some soul must at the moment of death fly the body and betake itself elsewhere. It is the most natural of all possible theories, even if it is not the current one. The dead body of even a Volvox seems suddenly to have been vacated. Something intangible seems to have departed from it. What, on the contrary, I do find surprising is not the assumption that men have souls, but that it should ever have come to be commonly assumed that no other creature has. The sense that something which was there is gone is almost as strong in the one case as in the other and nothing suggests that the death of one is radically different from the death of the other. The very word "inanimate" means "without a soul" and the fact that we still use it testifies to its appropriateness.

I assume that the biologists are right when they tell me that Volvox, having got as far as it did, seems to have got no farther. Perhaps some other creature independently paralleled his inventions — which would make the whole thing at least twice as remarkable.

Looking again at the Roller I console him thus: At least you were on the right track. Like many others in the history of invention, like, to take a very minor example, Langley and his airplane — you were on the right track even though something stopped you before ultimate success. Once you had transformed a colony into an integrated individual you showed how it was possible for living creatures to achieve more than merely microscopic size. Once you had invented the differentiation of the sexes you had started on the way to poetry as well as to rich variability.

It may also be true that none of your other inventions was more important than Death, without which none of the others could have been fully effective. As a certain multicellular, mortal and sexually-differentiated individual called Bernard Shaw has argued, nature would not have been able to experiment very freely with new forms if the earlier experiments were not removed after a reasonable time. The potentially immortal amoeba got nowhere. Only mortal creatures evolved.

D

3. The Animal's First Need

Is a VOLVOX any less remarkable than a bird, or a bird any less remarkable than a man? Of course it is — in a sense. But miracles cannot be compared. One is quite as incomprehensible as the other and if man did not exist a Volvox or a robin would be as difficult to "explain" as man himself.

To compare the three is like trying to compare infinites. You cannot say how many times greater one is than another because they are larger than anything that can be imagined. Everything that lives is incommensurate with everything that does not. It has characteristics no nonliving thing even hints at, and in that sense life is an absolute.

Of all these characteristics, the most indescribable is consciousness. To *be* and to know that one *is* is the ultimate privilege, and the ultimate burden, of Man. But whether or not consciousness is an essential and universal characteristic of living things we cannot possibly know. All that lives may or may not be at least dimly aware of itself. But whether it is or is not, there are other characteristics — like the ability to grow and to reproduce — which are possessed by the smallest, the simplest, and the humblest. And perhaps the most fundamental of all is the ineluctable necessity for food. Because the inanimate does not necessarily either change, or grow, or reproduce, it is self-sufficient. But even the potentially immortal amoeba must nourish itself or die.

Of Volvox, we admitted that we could not say decisively whether it was plant or animal, though between many of even the one-celled organisms the distinction is already quite definite. But what, after all, is this distinction which we so take for granted? Since we are to be concerned from now on almost exclusively with the animal rather than the plant, we should no doubt ask. And what the answer will finally come down to is: Food.

A delightful and still popular nonsense book is called *How to Tell the Birds from the Flowers*. This, you may say, is not usually very difficult. But if you generalize the question a little further, if you ask how to tell the animals from the plants, it is not so easy. In fact, biologists are usually hard-pressed when it comes to finding satisfactory definitions for such fundamental things or making fundamental distinctions in terms that will stick. In the case of Volvox they give up. As regards the higher creatures the best they have ever been able to do is to say: "Animals are compulsory protein feeders; plants are not."

This may seem pretty farfetched, or at least pretty irrelevant to what we have in mind when we think of an animal.

"I love animals" means something. "I love compulsory protein feeders" is nonsense. Nevertheless the definition really
is foolproof and it is the only one that is. "Compulsory" has
to be put in because of such insect-eating plants as the common sundew, the pitcher plant, and the Venus's-flytrap. All
of them consume protein though they can get along without
it. No animal can.

What this means in plain language is that all animals must
eat something which is or was alive. It may be either a plant
or another animal but only plant or animal matter contains
protein and without it they cannot live. No animal, therefore, can be innocent as a plant may be. The latter can turn
mere inorganic chemicals into living tissue; the animals
cannot. All of them must live off something else. And that,
perhaps, is the deepest meaning of Original Sin.

The soul — if there is any such thing — is unknown except
as it inhabits an animal body. And the animal body must be
nourished by what is, or has recently been, living a life of its
own. Hence the first necessity for every animal is the necessity of finding something to eat. He may or may not require
shelter to protect himself and he may or may not have to
discover or provide some sort of home for his offspring. But
find something to eat he must.

Of the temperamental Madam Fremstad it is said that
when she sat down to a certain dinner she flung the roast on
the floor with an indignant exclamation: "Pork before Parsi-
fal— Bah!" To a bird, on the other hand, worms are a
legitimate part of the joy of life and he sees no incongruity
in the fact that they are turned into song.

Most birds accomplish this necessary business in direct,

relatively simple ways. Nearly everybody has seen the robin tugging away at an earthworm who holds on for dear life — but usually in vain. The only mysterious part of the business is the robin's gift for knowing where the worms are and to this day a dispute still rages over the question whether he has been listening for his underground breakfast when he cocks' his head and then digs, or merely getting one eye into a position where it can look straight down.

The warblers search indefatigably for insects on leaves and blossoms; nighthawks and whippoorwills fly with huge wide-open mouths to funnel in gnats, flies, and moths on the wing. Your woodpecker makes the bark fly from dead limbs in order to get at what lurks beneath, though he also, especially in spring, drums vigorously just for the sake of making an interesting noise. And because he is a percussionist rather than a singer, he will be grateful for the novel timbre of a tin roof if you have one.

The sapsucker who drills little circles of holes around your apple tree to drink sap seems a bit more ingenious than most other birds, and certain woodpeckers who drive acorns into prepared recesses in the bark of a tree in such a way that the squirrels will not be able to get at them before the birds dig them out again in winter seem a little more so. Another woodpecker, one who uses a small stick for the same purpose, is astonishing. But on the whole birds tend to be mere hunters of small game or collectors of wild fruits who live off the country like primitive nomads.

Moreover, despite Western man's aversion to insects, their food seems reasonable and proper. But if every animal must eat something organic, it seems to be also true that everything organic can be eaten by something. Cockroaches thrive on the ancient paste beneath wallpaper and they as well as various other insects revel in the glue of book bindings. "Poisonous" is a meaningless term unless you specify "to whom." Thus the common chipmunk of the East nibbles with impunity the Deadly Amanita or Destroying Angel mushroom, one of the most potent poisons in the world. The so-called drugstore beetle can live for years sealed into a bottle of such violent poisons as aconite and belladonna.

Some creatures, including man, are almost omnivorous and in a pinch can be nourished on almost anything organic, though man cannot get any good out of the grass (not even if he cooks it) on which the ox grows great. Other animals are absurdly restricted, like the Australian koala "bear" who has caused zoo keepers endless trouble because he can't eat anything except mature eucalyptus leaves — young ones poison and will ultimately kill him. Most whales, including the species that grows to be the bulkiest and heaviest of all the animals of sea or land, nourish themselves preposterously on microscopic or nearly microscopic plankton floating in the ocean, though one sort — the killer whale — seems to go against nature by slaughtering almost any animal that comes his way.

Literally nothing is wasted. Everything nourishes something else until the bacteria finally get hold of it and return it to the soil after breaking it down once more into inorganic compounds which no animal could eat but which plants can again transform into protein. Hamlet shared the ancient delusion that decently buried men become the food of maggots, though actually "the devouring worm" cannot reach them. But he had, in general, the right idea when he exclaimed: "A man may fish with the worm that hath eat of a king, and eat of the fish that hath fed of that worm."

On the physical plane no doctrine could be truer than that of reincarnation. Every living body, including our own, has lived many times before. Humble plants seize upon merely chemical elements or compounds and organize them into more complex combinations which thereupon begin to live or, as some would say, to be used by life — which is not the same thing. The animal that eats them raises these compounds to even more advanced levels of complexity and they are inspired with more complex forms of life. The robin who visits his nestlings every five minutes is busy turning bugs into birds.

Omnivorous man may eat the animal and thus build his flesh from its flesh. If he is at last returned directly to earth then what was a man is resolved quite rapidly into plant food again. If his body is burned, then it returns by only one step to the merely inorganic. But if by any chance he is, as was once more common, devoured by some carnivorous beast, then at least one more step downward interposes. But by whatever stages he returns to chemical simplicity the upward phase will presently begin anew. After each fall there is a rise. During millions of years the level achieved at the end of the rise was higher than any that had ever been reached during any million years before man evolved. And nobody knows whether or not the upper limit has been reached.

Thus the robin pulling a worm from the earth or a whale straining plankton from the sea is not merely satisfying an animal need. He is lifting protein from one level to another. Perhaps, as some would say, the higher form of life is "nothing but" the more highly organized compounds that constitute the body of the bird, the whale, or the man. Perhaps, as others would insist, the more complex body is merely somehow necessary to the higher life which informs it. But in either case the individual could not start where the plant started. Some living thing, either plant or animal, had to manufacture for him those compounds that are the simplest he can use.

Only man seems to have made much of an art or ritual out of eating. And that fact is more surprising than might at first sight be supposed. After all he was long preceded by other creatures in the elaboration of various other necessary or useful activities. Thousands of years before he organized games for exercise, animals had been engaged in play markedly formalized. They had also elaborated family life and given to the business of caring for their offspring emotional concomitants that point plainly toward the human. Their ability to make sounds of some sort may have been at first merely a warning to enemies or even an involuntary cry of rage, but long before man learned to speak, animals had seized upon their simple gift and elaborated it into song or some other system of sounds ritually performed. As for the sex instinct, it began many eons ago to generate activities almost as protean as those of man himself. During courtship animals sing, dance, waft perfumes, display colors, and indulge in innumerable forms of mere showing off. In this department there was little left for human beings to invent.

But the art, the philosophy, and the ethics of eating are all almost his alone.

Among the lower vertebrates — frogs and snakes — eating is apparently a distasteful if not positively painful process, and some of them succumb to it only at long intervals. Predators, on the other hand, make it an occasion of savage fury. But few if any animals can be said to *dine* in the sense that they may be said to play, sing, make love, or practice the arts of homemaking and domesticity. And since, except for the domesticated animals, few will eat more than is good for them, it seems a reasonable conclusion that they eat to live but are hardly capable of living to eat.

Voltaire called man the only animal who drinks when he is not thirsty and he might have added, "Or eats when he is not hungry." Perhaps that is one of the reasons why he is also the only one who has developed in magic, in manners, or in religious dogma scruples and taboos in connection with eating — more of them indeed than in connection with any other natural process except the sexual. Thus the gourmet and the ascetic generate one another. To live to eat suggests, by reaction, that eating itself is sinful; and while the worldly try to see how much food they can consume, the monk not only tries to see how little he can survive on but regards the indispensability of that little as evidence of his depraved state. We think it indecent to eat much when we mourn, and when we celebrate we think it obligatory to provide more food than is good for us.

Some individuals as well as some civilizations have felt the impulse to compromise by abstaining at least from what Bernard Shaw calls "the practice of consuming the corpses

of animals," but the most solemn mystery of the Christian
religion (and it is not in this respect unique) is the moment
when we practice theophagy. The robin, nevertheless, is
neither ashamed of his worm nor inclined to elaborate either
the acts or the emotions which accompany its acquisition.
And what is true of him seems to be true of most animals.
Some birds do offer food as a demonstration of affection, but
that is as far as it goes.

The earliest men, like most of the so-called higher animals,
did not make a living; they simply found it. Like many animals they could hunt for food and perhaps store it up, but
they did not breed the animals they slaughtered and did not
plant or cultivate the nuts and berries they collected. No
vertebrate before Neolithic man was herdsman or gardener.

Perhaps this fact seems too expected to be worth stating.
And so it would be if a certain paradox which we will meet
again and again did not present itself. No *vertebrate animal*
is more than hunter and storer but many *lower* creatures are.
Many of the domestic arts had been elaborately evolved by
insects and become complex "cultures" millions of years
before man or any of his direct ancestors invented the simplest techniques to make his manner of getting a living very
different from that of the mere beast of prey.

Spiders spun silk long before even man himself could do
better than wrap himself in skins. Bees concentrated honey
from nectar long before he had even built a fire. For that
matter, even the birds who shared with him the technical
backwardness of the vertebrates built nests while he could
still do no more than seek shelter in ready-made caves.

So far as the arts and sciences are concerned, man was long what many of his fellow mammals still are: members of a very much retarded race. Whatever other characteristics he and they were developing these characteristics did not include any that resulted in more ingenious ways of living or of making a living. It must have been only a few thousand years ago that men developed a language capable of conveying any practically useful information as precise as that used by the bee when he tells his fellows how to find a good pasture he has discovered. And it was only yesterday that man invented navigational techniques as ingenious as those used by these same bees who steer by the polarized light from the sun.

If, as the anthropologists believe, man has been a social animal for not more than a million years then certain insects discovered the advantages of a cooperating group something like thirty million years before he did and our progenitors might profitably have heeded an injunction "go to the ant," not so much to learn industry as to learn the agriculture and the animal husbandry man had not yet dreamed of.

Nearly everybody knows that ants "keep cows." To your sorrow you may see, almost anywhere in the temperate zone, aphids overgrazing your favorite flowers and see the ant-dairymen milking their sweet juice. With a little patience you may even see the ants actually putting their milch cows out to pasture and you may curse them for their ingenuity. But it is somewhat less well known, though more astonishing, that other species of ants practice agriculture very elaborately, ingeniously, and successfully. No Paleolithic man about to reach the point where he would begin to plant

gardens and thereby qualify himself to be raised to the rank
of *neo* instead of *paleo* ever investigated the insect's method.
But if he had been capable of doing so he might have advanced a few thousand years at one jump on the basis of
what he could have learned about soil preparation and plant-
ing. The paradox of the socially retarded mammal ought to
be, but so far as I know has not been, investigated.

Most of the books, either technical or popular, which deal
with such things usually choose as their example of insect
technology one of the tropical agriculturists. But it is not
necessary to go outside the United States to find ant farmers
who have mastered very advanced techniques of agriculture.
The arid Southwest where I now live is the section of our
country where they have colonized most successfully. On
any late summer walk I am pretty sure to come across both
the communities of these farmers and other communities
that specialize in making a living in quite a different way.

In certain areas — especially those where there are a good
many clumps of a certain hard prickly grass which springs
up after a summer rain and is by now ripe and dry — I see
rings of yellow chaff surrounding the nest entrances, in and
out of which ants are streaming either with ripened grain or
with bits of chaff to be piled outside. This is a busy harvest
time and the ants seem to know very well what they are
doing. Solomon said not only, "Go to the ant, thou sluggard;
consider her ways, and be wise." He added that the ant
"gathered her food in the harvest." And Solomon was taken
to task by later naturalists who did not know that any ants
actually did harvest seeds and who either blamed him for

nature-faking or apologized for him by suggesting that the bit about harvesting was one of those convenient "late additions" to the ancient text.

It was not until 1829 that a British army man stationed in India first noticed and confirmed the fact that Solomon was in this instance a better naturalist than anyone in the centuries intervening since the heyday of his moralizing. But apparently even wise Solomon did not know that other species had made the great step from the mere harvesting of wild plants to genuine agriculture.

Neither his harvester ants nor mine are more than *mere* harvesters. They neither plant nor tend the grass from which they gather the grain. Theirs is a kind of operation which was, perhaps, not beyond the capacity of late Paleolithic man. But other ants have gone far beyond what either he or most ant species are capable of.

If I go into some area slightly more deserty than that which the harvesters favor, I will come across imposing piles of sand, six inches or more across, two or three inches high, and sloping inward toward the nest entrance so that the latter is at the bottom of a crater shaped like an inverted cone and hence designed to funnel any rain that may fall into the underground chambers. Instead of the harvesters' pile of chaff, scattered green leaves — often those of the resinous creosote bush — may be seen here and there.

These leaves do not look very nutritious, and for ants they would not be. Yet the ants are carrying them below ground as purposefully as their harvesting cousins carried the edible grass seeds. A broken trail of dropped leaves leads to a shrub perhaps fifteen feet away. Ants have climbed it, bitten off leaves, carried them to the ground and then into the nest. In fact they do all this on so large a scale that in tropical countries related species do enormous damage to orchards.

Neither these tropical species nor those of my desert are going to eat the leaves, though for many years it was assumed that they would. The leaves will be used indirectly. And this fact makes it all the more remarkable that these farmer ants should know their usefulness. It is, after all, one thing to recognize food and to take the next step of gathering it for future rather than for present use. Even a secondary use for such food might be accidentally discovered. But a great deal more foresight and insight of some sort, conscious, or by now at least completely unconscious, is required to gather something quite inedible which can nevertheless be used to produce food. Between the two there is all the difference between the nomad who gathers nuts to eat or to carry with him and the settled gardener who has learned that food can be grown, not merely found. And this actually is the difference between the harvesting ants and those who nip leaves from a creosote bush. Those leaves are to be used by gardeners who, in some sense, recognize their value as humus and as plant food.

I have never dug into one of their nests, but I have seen specimens of what you find if you do — amorphous dirt-brown masses of rather spongy-looking material one might be hard put to guess the nature of. Actually it is composed of more or less decayed leaves. And if it has been dug up at the right time it will be penetrated everywhere by the white threads of a fungus. This "mushroom" is the only food of the ants, who not only prepared their mushroom bed but planted the fungus itself where it would grow.

All the ants who have adopted this way of making a living belong to the New World; all belong also to a single tribe called *Atta;* and all the members of this tribe practice mushroom culture whether, like most of them, they live in the tropics, invade the Southwest, or, in the case of one small species, may get as far north as New Jersey. The first observers assumed that the leaves were eaten. As a matter of fact the account published by the first to grasp the truth was neglected and it was not until a German student in the eighteen-nineties elaborately described what he had discovered in the tropics that accurate knowledge of the agricultural ants begins. Since then many entomologists, including the great American student William Morton Wheeler, have worked out all the details and even suceeded in following the development of laboratory-raised specimens from egg to established community.

There is nothing hit-or-miss about the methods employed. Almost purely instinctive though the procedure is assumed to be — at least by now — it has all the appearance of being as intelligently purposeful as any of the steps taken by men to grow any difficult crop.

To begin with it must be remembered that mushrooms are a very difficult crop indeed. They require a very special soil and very rigidly controlled conditions of humidity and temperature. To preserve a pure culture or "stand" of a given kind is more difficult still because the air is full of fungus spores; where one kind flourishes others are very likely to gain a foothold. Yet each member of the Attid tribe does grow a special mushroom that furnishes its only food and does keep the culture pure — no "weeds" grow in their gardens. Moreover the crop is artificially controlled in such a way that it produces only the threads, which act like roots, and the tiny above-surface globules, which the ants eat. Remove the agriculturists from a nest and the garden soon goes wild, assuming a different vegetative form and producing less of the edible material.

Suppose you start your observations with a thriving colony of, say, the species upon whose aboveground activities I have been from time to time casting an eye — although I see nothing except the procession carrying creosote or mesquite leaves from the plants through the entrance leading to the nest. When Wheeler dug up a small colony of this very species he found that the chamber with garden was more than three feet below the surface.

Down there a group of workers are receiving the leaves, cleaning them carefully, and then chewing them into a pulp moistened with saliva. When a small pellet has been properly prepared it is added to the outer edge of the already growing garden, so that the garden becomes larger and larger as time goes on. Meanwhile other workers are keeping the growing mushrooms in good condition and feeding the newly hatched, wormlike larvae with bits of the mature mushrooms, which furnish the only food they or the adult will ever eat.

All this is remarkable enough. But how does a garden get started? Where do the spores to be planted come from?

To answer that question one must know first that most ants, like the honey bee, come in at least three (among the ants often more) different forms. There is the queen who lays eggs, there are sterile females who do most of the work ("maiden aunts," Thoreau called them), and there are the males who are good for nothing except that once in their lives some of them impregnate a virgin female.

The female ant leaves the nest only once. She and the young males both have wings and sometime during the summer, at what looks like a prearranged signal, the youths and maidens pour out of the nest, leap into the air, and begin a dance (What pipes and timbrels, what wild ecstasy!). I have not yet seen this happen at any of the nests of the mushroom-growers, but not long ago I did see the nuptial dance performed by the members of a colony of a certain less interesting red species.

From a distance you might mistake the dancing swarm for gnats, though the individuals are many times gnat size. Approach and you will see that mating couples, still joined, are dropping to the ground. Soon they separate and the life of the male is nearly over. But the female of the mushroom-growing species has many difficult things to do before she can become a queen in her own right and spend the rest of her life laying eggs that will be fertilized by the sperm she stored up during her one mating experience. Almost immediately she tears off the wings, for which she has not further use. Soon she will dig a little cavity and retire into it, never to emerge again into the light. There she will lay the eggs from which a generation of workers will develop to enlarge the nest, care for the next generation, and so on.

All this seems to involve the solution of many problems

even for the majority of ant species which collect food only where they can find it. But what of the Atta queen who must start a garden planted with a particular "seed" and begin to grow a difficult crop?

Well, the new Atta queen "foresaw" all the difficulties and made certain preparations, despite the fact that she had never before been outside the nest in which she was born.

The most important of these wise preparations was this: before she left on her wedding journey she bit off a piece of the fungus garden, and instead of swallowing it carried it as a sort of dowry at the back of her mouth cavity. Then, when she found herself widowed on her wedding day but safely underground in the tiny little excavation she had made, she removed the bit of fungus garden from her mouth, placed it carefully on the ground, and laid a few eggs beside it. Soon she must grow enough food to feed not only herself but also her young. There are yet no helpers to bring in leaves to fertilize the garden. What does she do?

What she does is rather indelicate, but not more so than what is involved in certain methods adopted by the wise Chinese. She fertilizes the incipient garden with her own excrement. And that also she does "deliberately." She breaks off minute fragments from the small growing mass, holds them under her anal vent, deposits a drop of liquid feces upon them, and then puts them back — near, but not on, the original tiny speck of culture from which they were torn. In this way she manages to raise just enough produce to feed the first hatched larvae. Because food is scarce they develop into stunted workers. But presently they are collecting leaves and soon the garden is thriving. As is so commonly the case with those who establish new colonies, the first year is the hardest.

Now how on earth did it ever come about that a mere insect had developed a highly complicated gardening technique untold millennia before man or any other vertebrate animal had taken the first blundering steps in even the general direction of agriculture? Henri Fabre would have answered: "Simply because God built the necessary system of instinctive actions into one of the smallest of his creatures." Most entomologists would reply almost as dogmatically: "Because in the course of millions of years accidental variation and blind chance resulted in useful actions which had sufficient survival value to fix the habits upon the race which happened to practice them and gradually to bring the techniques to perfection."

If you object that this seems improbable, they will reply that given time enough nothing is in this sense improbable. Everything that could happen will, in time, actually happen. Nevertheless, if it still seems to you that something more than blind chance helped the process along, you can make the orthodox account seem still more difficult to swallow by pointing out that the improbable thing must have happened not merely once, but three times. Some termites and at least one beetle also grow mushrooms. In both cases the techniques must have been developed quite independntly of one another and of the ants. Not once but three times agriculture was invented by an insect.

Wheeler, who recognized the difficulties, says finally: "These insects in the fierce struggle for existence, everywhere apparent in the tropics, have developed a complex of instinctive activities which enables them to draw upon an ever-present, inexhaustible food-supply through utilizing the foliage of plants as a substratum for the cultivation of edible fungi."

But Wheeler, who besides being a notable taxonomist and a very distinguished observer was also a man of considerable philosophical humor, must have realized that as *explanation* the sentence just quoted is on a par with Polonius' explanation of madness. For to define true evolution, what else is it but — evolving?

4. Parenthood

MOST PEOPLE are more interested in young animals than in grownups, and at any zoo the mother with her baby attracts the largest crowd. Parental concern is a touch of nature which even those usually indifferent to their fellow creatures recognize as making them kin. There is nothing else which suggests so strongly that the animal is living in an emotional world very much like our own.

Yet many animals get along without any parental care whatsoever and many parents are completely indifferent to their offspring. All except the very simplest one-celled creatures have to have a father and a mother — or at least a pair of great-grandparents somewhere in their past. But only the begetting is indispensable. Some are protected in youth and then educated; others are on their own from the moment of birth or hatching. And the difference is by no means always a difference between higher and lower or even a characteristic of the class to which a given creature belongs.

Most fishes and reptiles get little attention from their parents, yet the common stickleback of the ponds builds a nest faithfully guarded by the male. All young mammals must nurse, and to that minimum extent the mother must concern herself with them. But there is a vast difference between the minimum attention some must get along with and the long period of protection and of education accorded to others. Mammals receive less care than some birds and, with the exception of man, no other creature makes such elaborate preparation for the welfare of its offspring as many insects do.

In some sense the fox and the rabbit, like the cat and the bird, seem to love their young. On the other hand it can hardly be imagined that the toad can love the tadpoles it will never see. Yet when the toad returns to the water to lay its eggs it seems, from the outside, to be exercising a degree of foresight far beyond any that could be expected of any mammal other than man. And by comparison with many insects the frog does not do anything remarkable. The instinct upon which both depend enables them to perform feats only man, among creatures even partly dependent upon intelligence, can approach.

Just as the toad "knows" that its eggs must hatch in water, so the Monarch butterfly "knows" that the caterpillars which are its children will have to find themselves on a milkweed when they emerge; and the cabbage butterfly "knows" that its eggs must be laid on the leaves of some member of the cabbage family — even though neither butterfly has any special need for milkweed or cabbage at any other period of its adult life. Yet the butterfly will never see its children and would not recognize them if it did.

That there is a difference between the "wisdom" of the caterpillar and that of a bird is clearly implied by the difference between their conduct when something untoward takes place. Try to take the eggs from a robin's nest and it will attempt to drive you away, exhibiting unmistakable signs of distress. Scrape the eggs of a Monarch from the leaf of a milkweed and even though it has not yet finished the laying process, it will pay no attention whatsoever. It "knows" that they should be put on that plant. It does not "know" anything else about them.

Obviously, then, "mother love" sometimes accompanies parental care but is not an indispensable condition for it, and the paradox is this: there are two very successful ways of helping the young to survive. One is to be aware of their existence and to take more or less intelligent steps, as the cat or the bird or the squirrel does, to meet situations as they arise. The other is not to "know" in our sense what you are doing or how to modify an almost invariable routine. Yet from the standpoint of survival it is by no means certain that the second does not work quite as well as the first. Insects seem to survive rather more successfully than most mammals.

What then is mother love good for? A difficult question indeed. And one hard to answer on any of the biologist's usual assumptions, which seem to make the insect more "sensible" just because it has no sense.

Not all insects have, to be sure, developed any form of parental care. Take for example the common walkingstick of New England, which has at some time or other surprised most country dwellers when they have happened to see it.

No other insect of our temperate region has managed to perfect so successfully the sly device which consists in protecting itself from enemies by looking like something inedible. But it takes no care whatever of its eggs, much less of its children. The eggs are dropped casually to the ground, in such numbers that they sometimes patter off the leaves of trees with a sound like raindrops, and, as I know from experiment, these unprotected eggs sometimes lie two years before they hatch. Still, such carelessness is the exception rather than the rule. Pick out any six-legged creature and the chances are that he will do something remarkable when the time comes to prepare for that future generation which, in the majority of cases, it will never live to see.

Take for example the wasps — unpopular creatures, to be sure, but endowed with what looks like remarkable foresight and most admirably concerned with the welfare of their posterity. They are likely to appear as unwelcome visitors when fruit or sweets are put out at a picnic, and they are also often seen flying through the air with some caterpillar, some insect, or some spider clasped between their legs. The sweets they are consuming on the spot and it is their favorite food. Usually, at least, the living victim they are flying away with is not for themselves but for children who are not yet even eggs. It will be quite a while before these offspring will be ready to eat the game just taken by their thoughtful parents.

There are many kinds of wasps which follow many different customs. Some are sociable (among themselves) as bees also are; some are solitary. Some build the large familiar nest out of paper manufactured by chewing bits of wood into pulp; others prefer adobe structures. But like related tribes of human beings they have many "culture traits" in common, and the habit of laying up a provision of fresh meat for the exclusive use of their children is widespread among the "solitaries."

For our example we might as well choose the "mud dauber." Nearly everybody has seen its nest plastered against a wall in a barn, a garage, or even within the protected entryway of a country house, where it is still less welcome. Not everybody has seen the maker at work preparing the tube or stocking it with provisions. Still fewer have ever inquired what it is all about, though the story is one of those most often told in books of natural history.

This solitary mud dauber, unlike the social ants, or bees, or wasps, is not a creature whose whole life is a succession of incredible acts. He (or in this case she) was born as a wormlike grub, and without leaving the nest she made herself a cocoon from which she emerged as a fully grown winged insect. From then until the time came to prepare for the next generation she has led a simple life: feeding on flowers or fruits, sleeping where she happened to find herself, and assuming no responsibilities — unless you count as a responsibility being ready to receive a mate when he comes along at the proper time. But when her eggs begin to ripen she abandons her thoughtless life and exhibits what looks like remarkable foresight and remarkable solicitude.

First she collects, one after another, little pellets of mud — almost any sort will do if it is sufficiently damp and sticky. She adds them one to one until she has constructed several neat little tubes, each of which has required thirty or forty separate loads, and when they have dried she undertakes to lay up in them provisions for the children who are not yet even eggs. Just what kind of game she will seek depends upon the species to which she happens to belong. Some solitary wasps specialize in flies, others in beetles, others in caterpillars, still others in grasshoppers, cockroaches, or what not. But none ever deviates from its hereditary preference, and so the common New England mud dauber takes only spiders.

She may need as many as ten or fifteen for each infant and it will take assiduous searching to find enough suitable ones to provision a whole nest. But when she has located a victim she knows precisely how to handle it, and this is the most remarkable part of the whole business. She paralyzes it with a sting and then carries the victim to the mud nursery. Presently she will lay a number of eggs on the spiders and when the egg hatches the wormlike larva will start to feed upon the stunned but often still-living victim.

Our common mud dauber goes in for rather small spiders, which are easily handled, but some of her cousins are bolder — notably the large steel-blue and orange solitary of the Southwest, who attacks the great hairy tarantula much larger than herself and usually wins the duel with an adversary himself well supplied with both venom and the fangs with which to inject it. But our common New England species is no less remarkable than any of her relatives so far as the most surprising part of the performance is concerned. All have mastered the same extraordinary technique. The poison they inject through their sting into the victim penetrates one of its nerve ganglia; the prey is reduced to impotence, though still living, and is carried unresisting to the cell prepared for it.

Here we come face to face with one of the most hotly disputed incidents in all the vast repertory of insect marvels. It was Henri Fabre, perhaps the greatest of all modern observers, who first studied in detail the paralyzing technique, and he regarded his discoveries as the most remarkable of his career. The French species he selected happened to be a caterpillar-hunter. Fabre announced, first, that his wasp expertly sought out with its sting a particular ganglion of the victim's nervous system and, second, that the caterpillar was paralyzed but not killed, because its body must remain alive and fresh until the larva was ready to eat it. Wasps had no icebox but they nevertheless knew how to keep game from spoiling. Moreover, he thought, all this was one more proof of his favorite thesis: God must have built into the insects an automatic mechanism capable of automatically performing all the complicated acts they had never had the opportunity to learn and whose purpose they do not understand.

Since Fabre's time other observers have attempted to check his results, some because of an impartial desire to know, some perhaps because they wanted to reject his thesis. Some prefer to believe that insects did, at one stage of evolution, actually learn things for themselves. Others insist that on the contrary everything in their behavior is explicable on the basis of natural selection, which has gradually given the appearance of purpose to actions originally random. In any event the two Peckhams, who made more than half a century ago the classic observations on American solitary wasps, came to the conclusion that on this continent at least not all of Fabre's statements would stand up. The wasp does paralyze its victim. Often that victim does stay alive until it is eaten. But sometimes it dies and, so far as they could tell, a dead, partly decomposed spider was just as acceptable to the wasp larvae as a fresh one. Therefore, they were inclined to think, Fabre was overinterpreting when he maintained that the greatest marvel of all was the "foresight" exhibited by the hunter who paralyzed but did not kill his victim. Perhaps, they suggested, the purpose of the sting is merely to subdue the spider or caterpillar. Perhaps it does not matter whether it dies or lives on in a comatose state. Perhaps the fact that it often does live on is a mere accident.

This dispute does not, however, greatly concern us at the moment. Take the less wonderful version of the wasp's behavior and it is still wonderful enough and quite sufficient to establish our present point. Once in its life the wasp makes elaborate preparations for the survival of the offspring it will never see. Somehow or other the ability to repeat the performance in the future is transferred to the egg. Somehow or other this ability will be built into the larva and then, somehow or other, survive the almost complete destruction of the larva's body in the course of the sickness that will fall upon it when, without leaving the mud nest, it has spun the cocoon from which it will finally emerge as an adult wasp.

There are many insects whose preparations for the next generation are even more elaborate and occupy much more time. The social wasps, for instance, lay up no stores of living game because they or their firstborn are going to stay around the nest and bring fresh, carefully masticated food to feed the larvae as they grow. This is one of the things the adults are about when you see them swarming about the round paper nests in your apple tree or about the open combs attached to the roof of your porch. But the very fact that the mud dauber is only briefly seized by a concern for the future generation and that his other habits are relatively uncomplicated makes him the most convenient illustration of the thing we are concerned with, namely the fact that parental and, in this instance, purely prenatal care of a very effective and essential kind can be given by a creature as remote from us as the solitary wasp.

Here, surely, is an extreme example of doing something well without — so at least it would seem — having the slightest idea why you are doing it. And for that reason we can hardly feel that there is in the wasp's behavior any touch of nature which makes us kin. We may, like Fabre, see in his behavior the glory of God, or, like some evolutionists, a machine which seems to imitate purpose so cunningly as to arouse in some the suspicion that what is called purpose even in human beings is also only a simulacrum. Certainly, however, we would hesitate to talk about mother love in connection, with a solitary wasp. Yet we still have to admit that if it's survival value you are after then mother love doesn't work any better. Young wasps seem to survive just as abundantly as young foxes, or kittens, or human beings. And so we must ask again whether mere "survival value" can account for the existence in other animals of mother love as an emotion.

Because the wasps and all the other insects are presumably incapable of love or of any other emotion or thought, their techniques, no matter how ingenious or how successful, leave us cold.

The wasps are expert but alien. There is something about their blind foresight (an impossible but inescapable phrase) which is repulsive and terrifying. They belong to that part of the universe which operates beyond our comprehension and almost beyond our sympathy. They are precisely what we don't want to be. Theirs is a form of life which might exist forever and still be without significance for us. They go their way, we go ours. And between us there is no communication. We are capable of love; they presumably are not.

How different by comparison is the often blundering but conscious solicitude of warm-blooded animals. As the family cat approaches her term she searches restlessly for a suitable place to give birth to her babies — and often settles finally upon what seems to be a highly unsuitable one. As likely as not she will herself conclude that she has made a mistake and will move the helpless creatures to another, sometimes to another and another. The wasp was confident, unhesitant, and efficient. The cat is fussy, apprehensive, and uncertain. But in that uncertainty and fussiness we recognize something of ourselves. In the wasp we recognize nothing.

The wasp children will get no education and they will need none. Like us, the kittens will need it and like us they are more or less at the mercy of a mother's whims. She will wash their faces whether they need it or not and disregard all protests. She will bring them insects to play with and, later, a mouse. She will even sometimes snatch the mouse away again, acting exactly as the mother lion is said to act, and momentarily defend it with a growl just to teach her young that they must expect to face serious competition later on. When she takes them out of doors she will run along a gently sloping branch, encourage them to follow her, and then, on the sink-or-swim principle, try to nudge them off until they have learned to hold on with extruded claws.

What we find most engaging is not so much the wisdom of her actions as the apparent motives behind them. The very fact that, unlike the wasp, she does not quite know what to do about anything, the fact that her intentions are far beyond her competence helps us to recognize in her something akin to ourselves. And it may be even her greatest follies that seem most human.

Like human beings, some other warm-blooded creatures are quite capable of being so stirred up emotionally that they forget the occasion of their excitement and lose sight of their original intentions — as was plainly the case with a pet duck whose antics once amused me. Becoming a mother for the first time at the age of twelve years, she was frantically solicitous over the safety of her offspring. She had no trust even in the human beings who for twelve long years had never offered her anything but kindness, and she would fly at them in a fury if they approached too close. But her absurdity went far beyond that. If one persisted in coming near she would presently, in a very excess of anger, forget who it was she was mad at and start attacking the ducklings themselves.

Yet to say that this behavior revealed an impossible gulf between the mind of a duck and the mind of a man is to reveal an extraordinary ignorance of what the mind of man

F

is like. How often has some fond father, angry because something went wrong at his office, come home and punished for nothing the very children for whom his concern had made some unfavorable turn in his business so disturbing. Nothing is more human than striking at the object nearest at hand when under pressure of blind anger.

Konrad Lorenz, the great Austrian observer, tells the story of one of his ravens who fell in love with him. Not a man to consider personal inconvenience where there was something to be learned, he consented to receive the masticated worms which the affectionate bird insisted upon feeding him. But when the raven — a male — tried to entice him into a nesting hole he was physically unable to accept the invitation. Obviously the raven's insight into the situation was highly imperfect. Either he could not perceive, or was determined to disregard, the fact that a man is too large to enter a raven-sized nest.

No wasp would be capable of such folly — not because it has more insight, but because it does not need any. Perhaps, as many entomologists would be willing to grant, it is not the pure automaton Fabre tried to believe it. Possibly it is endowed with a dim consciousness and some power to adapt to a new situation. But it rarely depends upon either. The raven and the man are alike in that each depends much upon emotion, intelligence, and insight — all of which are, nevertheless, often inadequate to the situation.

A few weeks ago I happened to be walking along the sandy shore of a Pacific inlet where little snowy plovers were nesting. Several fluttered along the ground just out of my reach, putting on the broken-wing act that, as all hunters know, various birds have learned. Are the plovers aware or not aware that their ruse may serve to lead a predator away from the nest? Here, in other words, is an instance where it is impossible to say whether the act is blindly instinctive or accompanied by some understanding of its purpose.

Neither can we be sure in the case of the opossum or of the common, harmless little hog-nosed snake, both of whom "play dead" very convincingly. Mechanistically inclined biologists insist that they are merely paralyzed by fright and neither knows what he is doing, nor can avoid doing it. Indeed one experimenter has recently claimed that the possum can be "cured" of his tendency to hysterical fright by shock treatments like those now administered to human victims of certain mental disorders! But there is really no way of being sure how much, if at all, the plover or the snake or the possum knows the usefulness of his behavior. All that we can be sure of is that certain useful actions can be performed either with or without intellectual awareness and emotional involvement.

Field observers who have made the all-too-few classic studies of the higher mammals in freedom have shown conclusively that some of them not only protect their young but educate them in ways clearly implying a kind of awareness related to our own. In England, for instance, Tregarthen found the otter a most competent as well as a most solicitous mother. She plays with her children and also punishes their bad behavior. Unlike insects, otter babies are not born with a set of instincts to guide them along the paths their species should tread. They have to be taught to swim, to catch fish and frogs and rabbits. There seems little doubt that instruction is given and manners are learned.

Yet otters never become any more perfect in the art of

being successful otters than the orphaned wasps do in the art of being wasps. Instinct works more surely than either habit or intelligence. Yet we call the otter a "higher" animal than the wasp because the way in which it goes about the business of living is more like ours. It has intellectual awareness and, what is perhaps even more important, its actions are accompanied by emotions, whereas the awareness and the emotions of an insect must be exceedingly dim, if they exist at all.

An anecdote told originally by the English naturalist Hammerton has often been repeated by those who would dismiss as foolish sentimentality any concern over the emotional distress of even the higher animals. It seems that a certain cow of his acquaintance was so grief-stricken by the death of her calf that she refused to eat or give milk. The calf was skinned, stuffed with hay, and given back to her. She licked it affectionately, regained her appetite, and again gave milk. But the best was yet to come. Finally she wore a hole in the calf's skin, and when the hay came out she munched it contentedly.

Now the cow, like most animals which have been protected by man but not accepted into the human family as the dog and cat have, is no doubt a very stupid animal. But that is not the point, since it is the reality of the emotion, not the presence or absence of any clear insight into the circumstances surrounding it, with which we are concerned. The cow's intellectual understanding of the situation was obviously unbelievably dim. Her distress in the face of a mystery, nevertheless, was real and might be compared with that of a man to whom the sense that there is tragedy at the heart of the universe is as overwhelming as his inability to grasp intellectually the cause or meaning of the situation.

An experiment biologists are fond of citing is very relevant here. They say — I have never tried the experiment — that if you snip off the long slender abdomen of a dragonfly and present it to the mutilated creature's mouth it will eat this half of its own body with complete unconcern, and undoubtedly it would take quite as gladly the body of its mate or its child. Faced with its natural food, insect and cow alike did the natural thing — which was to eat; and the cow was as incapable of being put off by the fact that the food seemed to come from an unnatural source as the insect is of being put off by the fact that it has just been mutilated. But there is no reason to suppose that the dragonfly would be capable, as the cow was, of mourning the loss of its young. And that is what makes, from the human point of view, the immeasurable difference.

When, ever so long ago, the insect clan took the turning that led it down the road of ever-elaborating instincts and the very remote ancestors of the mammal took the other, which has led ultimately to man, each committed itself to momentous choices — one of which was, of course, between dependence upon a mind that could judge a situation and dependence upon a fixed pattern of behavior that could not easily be varied. But this is not by any means the whole story.

Somehow or other awareness means not only intellectual grasp but also emotional involvement. Both either first came into being or at least first became a conspicuous part of a living creature's existence a very long time ago, though not when life itself began. And from the human standpoint emotional involvement is quite as important as intellectual grasp. Even the animals with whom we live most intimately, the dog and the cat, bewilder us when we try to understand their minds. They seem sometimes so intelligent, so understanding; at other times so incapable of grasping a situation that seems to us overwhelmingly obvious. We never know quite what to make of them when we consider them as intellectually our kin. Often we wonder whether in our sense they can think at all, and a great gulf opens between us. But it is clear enough that they share our emotions even though they cannot share our thoughts. And it is not merely that they are glad or sad. We see them also jealous, hurt, sometimes ashamed. And here again the touch of nature which makes us kin is not intellectual but emotional.

Where did it come from when we find it either in them or in ourselves? If it seems to have less "survival value" than the insect's efficient instinct, then how can any "survival of the fittest" explain it?

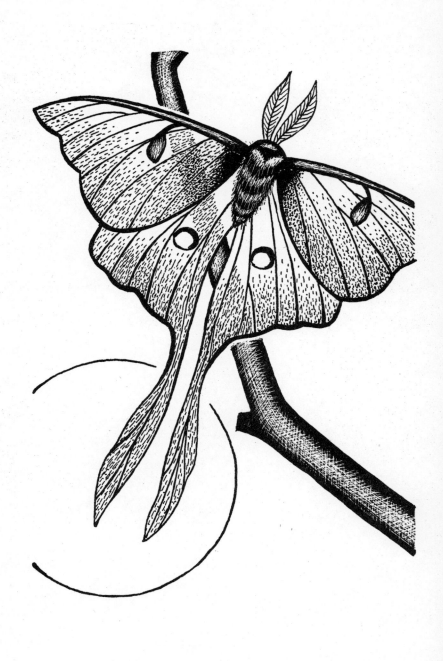

5. The Need for Continuity

AT THE BEGINNING of the fifth chapter of *Alice in Wonderland* Alice has an important conversation with a caterpillar. Thinking of her own recent experiences, she complains that it is very confusing to change size and shape. The Caterpillar — brusque as all Wonderland creatures are — replies: "It isn't."

"Well, perhaps you haven't found it so yet," said Alice; "but when you have to turn into a chrysalis — you will some day, you know — and then after that into a butterfly, I should think you'll feel it a little queer, won't you?"

"Not a bit," said the Caterpillar.

"Well, perhaps your feelings may be different," said Alice; "All I know is, it would feel very queer to *me*."

"You!" said the Caterpillar contemptuously, "Who are *you?*"

Having reached this impasse they proceed to explore

several others that do not concern us. But this one does. Like all other mammals we human beings take the continuity of our corporeal forms for granted. Except in Wonderland babies do not turn into pigs, or vice versa. If they did we might never have made the important assumption that our souls, our *personae*, or our egos, are similarly continuous. And if that assumption did not exist there would be no basis for that whole universe of ethical ideas without which men would not be men. Who would dare hold a butterfly responsible for what he did as a caterpillar?

The fact remains that for a vast number of all the different kinds of animal creatures on the earth today at least two bodies and two lives — sometimes several — are taken as a matter of course. And in very many cases neither the two bodies nor the two lives resemble one another in any way whatsoever.

The legless, vegetarian, gill-breathing tadpole grows legs, develops lungs, completely reorganizes its digestive tract, and crawls out onto the land to live the rest of its life as a meat-eating toad who never need go near the water again until something — it can hardly be memory — tells it to lay eggs in the water from which it came.

Who that has seen an ethereal Luna moth fluttering his great delicate wings on the windowpane and looking as though he had indeed come from the moon would ever guess that he was only a few weeks before the sluggish but ravenous green worm so repulsive to most people? Or that the harmless lacy-winged antlion, who looks so much like a miniature dragonfly, was once a flat dark-colored little bug lying hidden at the bottom of a sand pit waiting to seize with murderous pincers the ant who tumbled into his treacherous trap? The wisest child of such parents cannot possibly know his own father — or for that matter the father his own son. Neither could any human being guess which children came from which adults. To this day it sometimes turns out that some creature long ago baptized by science with a specific name is merely the young form of some other separately named and classified.

Even with us, children go through a sufficiently bewildering experience in growing up. They become aware of desires unknown before and they are often painfully embarrassed by minor physical changes, such as the breaking of the voice or the swelling of the breasts. Yet by comparison to the life of a frog or a butterfly the corporeal development of a mammal is extremely uneventful and unimaginative. If variety is the spice of life we have very little of it, and we make such a fuss over this little that it is hard to imagine what the problems of adolescence would be if young people fell into a deep sleep at fourteen and then came to with wings.

Your butterfly has to be literally born again. He returns to something like an embryo and then he grows up different — almost as though he were correcting some early mistake.

What, aside from the excitement of living two lives, is the point of that? What does the insect gain, and how in the course of what we glibly accept as "evolution" did he ever develop such a design for living?

Even the most self-confident biologist is likely to answer these questions less readily than he does most others, and it would be a pity to take even the best answers before comprehending fully the phenomenon. What visibly happens is something one would never believe if one did not see it. And the physiological process behind the visible happenings is no less remarkable. Before we even raise the question why, let us consider what does take place.

Many a lover of nature, many a professional biologist even, continues his boyhood habit of collecting a few larvae or chrysalises in summer and of tending them until the day comes when a dead worm bursts out of his mummy case to flit away on wings his humble form never seemed to promise. "Creeping" and "flying" are established in our vocabulary as the very symbols of the most contemptible and the most glorious forms of motion. Yet the caterpillar who went to sleep ignorant of anything except creeping wakes up to waft himself nonchalantly away on the most beautiful wings either nature or the human imagination has ever been able to imagine. No wonder that even the Greeks, un-otherworldly as they were, could not help being led to think of an airy soul leaving the gross dead body behind. "Psyche," or "butterfly," was their word for "soul."

Suppose we select for observation one of the most familiar and widespread of American butterflies — the large red-brown and black Monarch, which few dwellers even in the depths of cities have missed seeing. It ranges over the entire United States and in recent years has migrated, perhaps on American ships, to England, to Australia, and the Philippines. Country dwellers often see Monarchs gathered into large flocks in late summer ready to begin a southward migration very surprising for a butterfly, because most kinds either perish, leaving only eggs or chrysalises behind or, in some species, hide the winter out almost motionless in some shelter.

During the summer each female Monarch had sought out some member of the milkweed family and glued to its leaves a number of tiny eggs. Seen under a hand lens they are distinctly pretty — greenish in color, conical in shape, and neatly ribbed — but they are also too small to be noticed often except by those who look for them. Presently the egg hatches into a tiny caterpillar, which immediately begins the only business of its young life — immoderate eating. If the pasture holds out it never leaves the plant upon which it was hatched and does nothing besides eat, except on the several occasions when it pauses briefly to shed the skin grown too tight to be longer endured.

Presently it reaches full growth and it is then that the casual country walker usually notices it first. It is fat, soft, and to most people repulsive. But it continues to eat ravenously, so that sections of the leaf disappear visibly as the creature moves its head up and down the edge, taking great bites as it goes. Cylindrical feces almost as big around as the caterpillar itself fall to the ground one after another in rapid succession and do not mitigate the general impression of grossness. Moreover the caterpillar has made itself as conspicuous as possible by ringing the green of its body with black and yellow stripes. Monarch larvae have a bitter taste detested by birds and every member of the race is making sure that any inexperienced bird which tries one will not find it difficult to remember what a Monarch larva looks like.

Pick him up (if you are not too squeamish) and you will see that, crowded together near the front end, are the six

normal-looking legs all butterfly larvae have which correspond to the obligatory six legs of all adult insects. But there
is nothing else about him remotely suggesting a butterfly or
indeed an insect of any kind. These six close-together legs
are not sufficient for his length, and yet there is an absolute rule of nature that no butterfly can have even in
the larval state more than six. You will therefore see at the
hind end a double row of fleshy little stumps that clasp the
leaf upon which the creature is feeding and serve as substi-
tutes for legs.

If you decide to raise him in captivity for purposes of
observation and should this be your first experience, then
you will presently be filled with despair. Your caterpillar
seems sick. He has grown very sluggish and, incredibly, his
incredible appetite fails, no matter how fresh the leaves of
his favorite — in fact his only — food you may provide for
him. But do not be alarmed. Your caterpillar is about to
lose his life in order that he may gain it. Beneath his skin,
invisible to any observer, drastic changes have been going
on. He no longer eats because he could no longer digest if
he did. Moreover, a tough membrane enclosing his whole
body is forming just beneath the skin and it is making him
so stiff that he can no longer move freely.

Soon he will take up a position either on a twig of the food
plant or perhaps upon some other nearby support, natural
or man-made. From a special gland at his rear end a little
gluey substance will be secreted, and as it hardens it will
attach the tip of his body firmly to the support. No butterfly
caterpillar spins enough silk to wrap himself, as many moths
do, within a cocoon; but the Monarch glue is liquid silk.

Soon after he is firmly attached he will let go with all his legs and hang head downward. This is the last step before the caterpillar becomes a chrysalis.

The transformation from sick caterpillar to quiescent chrysalis is less often observed than the emergence of the adult because it happens so quickly. As a matter of fact there isn't much to see. Everything except the one final event takes place beneath the outwardly unchanged skin. Suddenly this skin splits near the rear end and, tissue-thin, drops to the ground. Where a moment ago there was only a worm there is now the mysterious chrysalis. But the Monarch was a good choice for observation because his chrysalis is the prettiest of any American butterfly.

The color is a pale, luminous, leaf-green; the shape somewhat ovoid but broken gracefully into two sections differently curved and as right as an amphora. Probably it follows "the laws of dynamic symmetry" (if there are any such things). It is chastely ornamented with a row of small gilded dots raised slightly above the surface where they half encircle the chrysalis, just at the line where the curvature changes.

"A green coffin with golden nails" someone once called it, and the phrase is accurate as such pretty phrases usually are not. The dots really are not yellow but precisely the color of gold leaf. If they serve any practical function I have never heard it suggested what that function might be. They can hardly be explained by any of the methods commonly used to reduce the beautiful in nature to the merely utilitarian. There is no "sexual selection" to be made at that phase of the Monarch's life cycle. They are certainly not conspicuous enough to serve, like the caterpillar's stripes, as the warning "I taste bad." Nor can the gilding be explained away as the necessary mechanical result of structure as the beautiful designs on the invisible diatom sometimes are. Possibly some substance necessary to the caterpillar but no longer usable just happens to be golden and just happens to be excreted along the line so gracefully placed in relation to the design of the whole. At least we will let it go at that.

Keep the green coffin under observation for a while and presently the yellow-brown of the developing wings will be seen through the thin transparent outer skin. Then one day the skin will rupture and a rather sorry-looking creature, rumpled and feeble, will somehow catch hold of a support with its six legs and rest motionless for several hours. Its abdomen is disproportionately fat, its wings crumpled into little disorderly packets like the not-yet-unfolded petals of a filmy poppy. Gradually they expand and pass from shape-lessness to shape. As they do so the fat body grows slenderer, because the fluid that once distended it has been forced through the veins of the wings to expand and stiffen them.

As they grow, the whole insect begins to gain in strength and confidence. An hour ago it had been almost as helpless as a premature baby and hardly able to cling to its support. In an hour more the wings will begin to flutter and the Monarch will sail away — not to the very brief life of many moths, but perhaps to flit until the end of summer and then, with crowds of companions, to start the migration that will carry it hundreds of miles away to some southern clime. Next spring a few with battered wings will make their way north again. Are these northward bound individuals retracing their previous journey, or are they members of a generation begot by those who went south? The question has been long in dispute. Butterflies cannot be banded as easily as birds!

In any event, few other things looking so fragile stand up for so long under the buffetings of life in this world. In so far as it is a machine, it runs not on gasoline but on the energy supplied by the sugar it gets when it sips the nectar of flowers with the long tongue that can probe so deep. Perhaps, in so far as it is more than a machine, it runs on its own tiny portion of whatever else it is that makes man and beast determined to live.

Anyone who has thus watched the progress from caterpillar to finished butterfly has seen incredible things happen. If he has any capacity for wonder he may have been too stunned to ask any questions or to want to probe any deeper. But sooner or later it may occur to him that a great deal has been kept hidden, that what he has seen is merely a series of suddenly revealed transformations. The skin of the larva split to reveal a chrysalis already formed. The skin of the chrysalis split to release the still crumpled but fully formed adult. For all he has been able to see, the two transformations were almost as sudden as the metamorphoses of the fairy tale. Some enchanter waves a wand and a man is a beast or a beast is a man. Yet we know that nature does not work that way. Nothing grows except through a series of steps. And these steps have been hidden from the observer. Twice the veil was rent but each time he was presented with a *fait accompli*. What actually went on within the coffin

G

with the golden nails? By what steps does nature make a caterpillar into a butterfly?

Those questions are not wholly unanswerable — at least in the same limited sense that we can answer how nature makes a chicken out of an egg. And the answer is similar though somewhat more complicated, because in this case nature has first to make an egg (or at least an embryo) out of a chicken. When our caterpillar looked sick he really *was* sick. He had been struck with what might well be mistaken for some sort of suppurating infection. His muscles and his organs were beginning to dissolve into what looked like a sort of pus. He had just life enough left to perform the final rites which accompany his entombment within the chrysalis before he was returned almost to death.

If we had been watching a fly — a common housefly for instance — the degenerative process would have continued until the maggot had been all but completely dissolved into a creamy liquid seemingly as featureless as the yoke of an egg. Had we been watching instead one of the presumably primitive insects in which the adult is not as different from the infant as is the case with the fly or the butterfly, comparatively little destruction of the already organized living material would have taken place. But here in the case of our Monarch the amount of degeneration which must occur before the larva can be born again — and born better — is between the two extremes.

Once the skin has been shed and the creature — or what is still left of it — ceases to move, the destruction of the original organs and the fashioning of new ones goes on apace. Free-moving cells, much like the phagocytes or white blood corpuscles in the human body, absorb and carry away the disintegrating material, and at the same time new organs begin to form. Ever since the day when the caterpillar was hatched from the egg it has carried within its body certain little groups of cells which were useless until now. They are the buds, if the term be permitted, from which the butterfly's organs will develop, and these organs grow on the material the phagocytes have been carrying away from the parts of its dead self. From these buds a butterfly began to form as the caterpillar was dissolved. No new material save perhaps air and moisure is available any more than anything is available to an egg closed within its shell. What is more remarkable, perhaps, is that almost nothing is left over. The material in one caterpillar is just sufficient to make one butterfly!

"Almost nothing" — the qualification is interesting. If you have kept your butterfly under continuous observation from the time it ruptured the skin of the chrysalis until the moment when it took wing you probably observed, at some moment not long before the last, that one or two drops of liquid fell from its rear end to the ground. "What," you may have asked, "is that?" It was Nature's miscalculation; or rather, her margin of safety. Being sure to have enough — and she cannot predict just how much was going to be lost by evaporation from the chrysalis — she had one or two drops left over.

Different phenomena strike different people with sudden amazement. And it so happens that the drop or two which fall from the rear end of a newly emerged butterfly has long seemed to me among the most astounding things I have ever observed. Whatever told the creature so much in the course of its already eventful life·tells it one thing more: "You won't need that now."

The legends of many peoples are full of changing-into-something-else not unlike what the butterfly takes for granted. Ovid's *Metamorphoses* — by no means an entomological work — is one of the most enduringly popular books ever written. The impossible possibility that a man or even a beast might turn into some wholly different creature seems to fascinate something buried deep in human nature. Did it arise, one wonders, independently as one of the never-to-be-fulfilled dreams of which presumably only man is capable, or was it suggested to him by the fact that so many other creatures do change out of all recognition? The question is not likely ever to be answered, but if it could be it might throw some light on another important one: Is what we call "the imagination" limited — as an old theory of psychology held that it was — to the recombination of materials supplied it by the senses, or is it capable of genuine creation; is it able to body forth before the mind's eye what never was even in its constituent parts; what, in some cases, never could be?

But why and how did so many creatures acquire the power to turn and the habit of turning into something else, when that is the very last thing possible to so many others? It can hardly be just because some feel more strongly than others that variety is the spice of life. Or at least that is not the kind of explanation to satisfy our habits of mind. We look for and we often find "explanations" in terms of mechanism and function which are temporarily satisfying. And though in the case of insect metamorphosis they are not so pat as they are in certain others, they do exist.

Aristotle knew that a butterfly was first a worm and then a chrysalis. He describes the sequence briefly in that great compendium of information and misinformation called the *History of Animals*. It was certainly part of the current lore which he summarized and there is no knowing for how long some men at least had been familiar with the phenomenon. But Aristotle says nothing about the how or the why and even today biologists are likely to be unusually tentative in their discussion. Nevertheless, ever since the theory of evolution inspired them with the hope of explaining everything in evolutionary terms, the problem has been repeatedly approached.

The most confident statement which most would be willing to make is a negative one. What was once generally supposed to be one of the master keys won't unlock this particular secret. Whatever the explanation may be, it is not that a modern butterfly has to be a caterpillar first because it was once nothing but a caterpillar and must therefore, in the course of its development, go through its evolutionary history. It is not, in other words, because "Ontogeny recapitulates phylogeny," or, in plain English, because the development of an individual briefly summarizes the evolutionary development of the species to which it belongs.

That may explain why, for instance, a human embryo has gill slits like those of an embryo fish and has also at a later stage of embryonic development a good deal more of a tail than the miserable, too easily broken little projection at the end of the spinal column which is all the adult can boast of to represent the imposing appendage his monkey-like ancestors found useful. It may even explain why children like to play Indian at a certain stage in their lives. Perhaps at that age they really are Indians. But it won't explain what happens to a Monarch butterfly. And there are many reasons why it won't.

In the first place, when ontogeny really does recapitulate phylogeny — and the rule is not as invariable as was once supposed — it doesn't do more than suggest rather than realize the earlier structure of the species. The gill slits of the human embryo never become functional; the future human being never breathes water with them. And if the caterpillar were a mere recapitulation of something in the racial history of butterflies it wouldn't be the perfect functioning organism it is.

In the second place, the fossil record of insects forbids the assumption that the immediate ancestors of the butterflies were caterpillars in the same sense that the immediate ancestors of the seals were land-dwelling creatures and the immediate ancestors of toads were water animals which later took to dry land. The fact that seals return to the land to give birth to young which have to be taught to swim and the fact that toads, on the other hand, return to the water to lay the eggs which hatch into water-breathing tadpoles really does illustrate a sort of recapitulation and the fossil record confirms what the habits would suggest.

But there is nothing to correspond to all this in the case of the butterfly. The oldest fossil insects belong in the class with those modern kinds which hatch out as something very like miniature adults and undergo only minor changes. As the paleontologist follows the story forward through time he finds that as more and more "modern" types appear the changes taking place in the course of the individual's life history become more and more radical. In other words the complete metamorphosis, instead of being something brought down from the very remote past, is something developed fairly late in the course of evolution. In fact it seems pretty safe to say that the modern caterpillar, the modern chrysalis, and the modern adult butterfly have all evolved into their present complexity more or less simultaneously, not one after the other. Hence the caterpillar as well as the winged adult are both "modern." Of course it may be true that all animals, including man, go back ultimately to the famous "wormlike creature." But the butterfly did not come from it much more directly than the mammal did and the caterpillar is not a mere intermediate stage between the two.

This leaves us with a fact almost as staggering as the fact of evolution itself. We have to recognize not only that life began somehow and slowly developed the orderly processes by which an egg becomes a child and the child grows up into an adult but that somewhere along the line certain living creatures also developed the power, not implicit in the other system of development, to change radically their whole structure and all their habits at the midway of their mortal life: the power to start out as caterpillars and to end up as butterflies. This fact the biologist simply has to accept, and if he wants to go on asking questions he must turn to those less ultimate in their implications.

Granting the fact that metamorphosis did arrive relatively late, the layman (or even the neo-Lamarckian) might be tempted to put such a question in the form: "Well then, why should any creature have *wanted* to live more lives than one, to submit itself to so drastic a break in the continuity of even its bodily structure and its habits of life?" A more orthodox formulation would be: "What purpose is served? What advantage in 'the struggle for life' does such a course of development confer?"

You can say, if you like, that all insects have their skeletons on the outside and that therefore they can grow only if they molt these skeletons from time to time. You can add that for some reason the skin or skeleton of functional wings is never molted so that flying insects can't grow once they have become capable of flying. Ergo, butterflies must be something other than butterflies until they have reached their maximum size.

Rather more simply, you can also argue that when winter or some other unfavorable season has to be endured there is an advantage in passing it during a resting period like that of the hibernating woodchuck or an erstwhile caterpillar now withdrawn into its chrysalis. Or you can say that certain advantages may result from a sort of division of labor which lets the caterpillar devote most of his time to accumulating nourishment while the butterfly is left free to concentrate upon the business of mating and egg laying. But it has to be admitted that many creatures manage quite successfully to survive the winter without turning into something else and to eat as well as make love during the same life stage. The explanations are satisfactory only in the limited sense that it is satisfactory to say that a nighthawk has a large mouth so it can catch insects on the wing and a woodpecker a long bill so that it can dig them out of dead trees. Either method works and we still don't know why one was adopted in one case, the other in another. Perhaps there is no reason except that more different kinds of creatures can be supported on this earth if every possible opportunity for making a living is exploited. Nature abhors monotony as much as she abhors a vacuum.

No doubt it would be very exciting to a human child if he could look forward to sprouting wings somewhere well this side of the grave. Still there are compensating advantages in not being a butterfly, and one of them is that we don't have to sacrifice continuity: that mentally and spiritually the human child is father to the man, that even the kitten is father to the cat, in a way that a caterpillar cannot possibly be father to a butterfly.

Notoriously the insects, even if not quite the pure machines they are sometimes called, learn precious little in the course of their lives and depend very largely upon inherited instincts to solve the problems of living. Even the "highest" — the ants and the bees — can profit extremely little from experience. Nearly all the "progress" evolution has made possible for them is in the direction of more and more complex patterns of instinctive behavior, not in the direction of increased intelligence as we understand the term.

It may very well be that the discontinuity in their lives is at least partly responsible for this fact. Even in a caterpillar inherited instincts are, to be sure, somehow carried safely through that near-destruction of his organization which takes place in the chrysalis. At least the butterfly "knows" that its eggs must be laid on the food plant which the larva, not the adult, must have. But it is hard to imagine that anything *learned* by an individual could similarly survive.

Once, many millions of years ago, the insects were by all odds the most up-and-coming of living creatures. They had a long head start on all the rest of us, and some think that in certain respects — notably in respect to social organization — they are ahead of us still. But if you measure progress in terms of consciousness, adaptability, or the power to learn from experience, then radically different creatures passed them long ago.

It has often been pointed out that the line the primitive ancestors of the insects first took imposes certain limits upon future development. It worked and it still works amazingly well within the limits, but the limits cannot be passed. If, for instance, you are going to have your skeleton on the outside and get oxygen through a simple system of ramifying tubes, then you can't get very big and a man-sized insect is mechanically unthinkable.

At first sight the vertebrate scheme — an internal skeleton exposing the soft vital parts to all the dangers of any physical environment — may seem to be much less advantageous. But the compensating advantages have proved overwhelming. One of them is that you can grow without molting. And if it really is because insects molt that they have developed the habit of metamorphosis then that habit itself may be responsible for their failure to develop what we call minds and for their consequent dependence upon instinct. Your butterfly does not have to be educated. He is born knowing everything he will ever need to know. But that means also nearly everything he ever can know.

All things considered, it is perhaps just as well that human adolescents don't retire into a chrysalis at fourteen. Forgetting everything they had learned up to that age would be too large a price to pay even for wings.

6. The Barbarian Mammal

THE ANT practices his incredible agriculture almost at my doorstep. The caterpillar will be born again on my study table if I put him there. Yet both of them live in a universe so remote from mine that they are not aware of even physical propinquity and they go about their business as though I did not exist.

So far as mere "behavior" — that idol of the laboratory — is concerned I can on the other hand know much more about them than about the "higher" mammals who are with me in my own psychic universe. I can bring a cat, a dog or some much less likely animal into the house, but either he will never lose his sense of strangeness or he will lose it so completely that he lives my life as much as his. What can I know of the fox and the deer and the badger, who survive even in suburban areas, much less of the wolf and the bear and the cougar, who actually possess the woods into which it is not they but man who is an occasional intruder?

Actually I know from direct experience little more than the nearly nothing to be learned from the sight of a fox pouncing catlike upon mice two fields away, from the woodchuck who used to sneak down the hill to steal my apples, or from the deer who tosses his heels and his tail as he removes himself as rapidly as possible from my observation. At most I may come across the nest of a bird, the burrow of a rabbit, or, here in my present home, the form of a hare.

Until comparatively recently nobody knew much more and most took it for granted that there was not much more to be known. Even now when more systematic observations have been made there is no reason to suppose that the ways of the mammals are so complicated, so strange, so ingenious, or so marvelously successful as those of the insect, the amphibia, or even the fish. This may be in some part because, being aware of us as the other creatures are not, they are more difficult to observe intimately. But it is also certainly because there is not so much of one kind of thing to know. Outwardly, the lives of the large animals are relatively bare and simple.

There is, to be sure, more known now than was known a generation ago about a number of things: about, for example, the migration of the herd, or the defense of "home territory" by certain animals, and the looser defense of the larger "home range." More is also known about the simple hierarchy within groups of more or less social mammals where, for every individual, the group is divided into those whom the individual dominates and those who dominate it.

But what are such migrations by comparison to the great, orderly migrations of birds; or such hierarchies by comparison with the elaborate caste system in an anthill? For the most part mammals lack not only the kind of "culture" which "lower" creatures have developed but also what looks like any comparably complicated behavior patterns. They are often nomads, frequently have only temporary homes if any, and hardly carry cooperation beyond the elementary stage it has reached in a wolf pack.

The appeal these large warm-blooded animals make to the imagination is not, as in the case of the insects, based upon the paradoxical contrast between a culture analogous to ours and the incredible, inexplicable fact that the closeness of these analogies is contradicted by the remoteness of the insect's psychology and the mysteries of the inhuman road by which he reached what looks like human institutions. In most respects the capacities and talents and emotions which the mammals do possess are, on the other hand, simply much restricted versions of those which we, being fellow mammals, also possess.

Strangely, the wolf and the deer are primitive in a sense that the ant certainly is not. And it is in that primitiveness that their claim upon the imagination rests. It is much like the appeal of the red Indian or the Bushman. The wolf and the deer are Noble Savages, romantic embodiments of the ideal of freedom, and simplicity, and wildness.

> I am as free as Nature first made man,
> Ere the base laws of servitude began,
> When wild in woods the noble savage ran.

Someone once bestowed upon Henri Fabre the title "Insects' Homer," and for all its inappropriateness it stuck. Actually, insects are not susceptible of Homeric treatment.

Their lives are too complicated, too narrow, too fixed by convention and too often very unseemly. What they require, and what in Fabre they got, was less a Homer than a Zola, an Ibsen, or even a Strindberg. However, the American elk, the mountain sheep, and many another large mammal really does live a Homeric life, and it is in Homeric terms that they could best be celebrated.

Looked at too closely the society of the ant or the bee has suggested to many observers the ruthlessness and brutality of totalitarianism. Less closely examined it may suggest at least a parody of what we call in human beings the social virtues. But you would never be tempted to call an ant noble. Worthy, perhaps. But not noble. Yet "noble" in just the sense that primitive human beings are "noble" is precisely what the lives of many large mammals do suggest. Culturally they are as little advanced as the Homeric heroes, but that is one of the reasons why their lives seem epic.

Their virtues are courage, daring, and the fight to the death. Like Homeric heroes they usually claim wives by violence and as is proper to the epic, the passion of love is little elaborated. Birds court like troubadours and the sexual mores of some insects are so *outré* that one hesitates whether to call them merely immeasurably ancient or decadent and perverse. But the large mammals are more likely to fight for a wife than to court her and when the polygamous bighorn or the seal fights a duel it seems as though Honor rather than the possession of any beloved is at stake.

To all rules there are exceptions and it would not do to say that solicitude and tenderness are never shown by males toward either their mates or their offspring. But it is certainly not among the large mammals that one looks for the most striking examples of domestic affection or romantic attachment. Hector may have said a tender farewell to Andromache and to little Astyanax. But to him the coming duel with Achilles was much more important.

Any one of a dozen of the so-called game animals would illustrate these facts. But no American mammal is more truly Homeric than the Rocky Mountain bighorn. His "culture" is neither that of the mere hunter nor that of the more advanced agriculturist. It is, like that of the Homeric hero, pastoral. The bighorn has no domestic arts such as even birds have developed. Shelter, when he must have it in winter storms, is merely where he finds it. But he has the strength, the courage, and the muscular skills of the hero. Moreover the glory of his life is in single combat. Like the Homeric hero again, he may make a dispute over wives the ostensible occasion for such a combat. But he scorns courtship and it is to the combat itself rather than to the occasion for it that he gives his heart.

Most living Americans have never seen this "noble" animal except in a national park or a zoo, and neither have I. Once he ranged in thousands over an area of some two million square miles stretching from Canada down into Mexico and he was still very numerous as late as the second half of the nineteenth century. Now he is threatened with extinction — as what large noble animal is not? — and he can be found truly wild only in the most inaccessible places, because he is no more a match for high-powered rifles with

H

telescopic sights than Achilles would have been. But he has had his poets and none perhaps better than John Muir, to whom he was the greatest of all the mountaineers: "Possessed of keen sight and scent, and strong limbs, he dwells secure amid the loftiest summits, leaping unscathed from crag to crag, up and down the fronts of giddy precipices, crossing foaming torrents and slopes of frozen snow, exposed to the wildest storms, yet maintaining a brave, warm beauty." From some such magnificent creature as this the domestic sheep have descended, but the heroic virtues cannot survive domestication.

A mature ram averages some three hundred pounds in weight, is covered with hair as well as wool, and bears proudly upon his head the magnificent curling horns fifteen inches in girth at the base, three feet long around the outside curve. "Besides these differences in size," Muir continues, "color, hair, etc. . . . we may observe that the domestic sheep, in a general way, is expressionless, like a dull bundle of something half alive, while the wild is elegant and graceful as a deer, every movement manifesting admirable strength and character. The tame is timid; the wild is bold. The tame is always more or less ruffled and dirty; while the wild is as smooth and clean as the flowers of his mountain pastures."

Never solitary and yet only somewhat gregarious rather than truly social, the bighorn wanders in smallish bands in search of pasture. He was not always confined as now he is to inaccessible mountainsides, for it was man who drove him for safety out of the less rugged valleys where he once spent much of his time. But he must always have been an incomparable mountaineer — not a climber like the goat but, as Ernest Thompson Seton says, "a sure-footed bounder like the Chamois."

Impossible stories are told of his plunging heedlessly from precipices hundreds of feet high or of leaping up to the top of others scarcely less lofty. That is an illusion. Actually he cannot leap more than six or seven feet straight up or broad-jump for more than a generous fifteen. But he is incredibly sure-footed: he can bound to narrow ledges, from which he again takes off; and in that way he negotiates cliffs which would be certain death, not only to man but to most other "sure-footed" animals.

In discussing these wonderful feats, Thompson Seton says:

One should begin by considering the weapons with which they are done. First, the quick calm brain, the "strong heart" as the Indian would call it; second, the superb outfit of tireless, tense, life-tingled muscles; third, the tremendously strong and supple feet, equipped each with two great soft rubber pads that can grip on any surface, hard or soft, rough or slippery. Cliffs that are perfectly smooth and plumb are so rare as to be negligible. No high rock overhangs for any considerable distance; gravity forbids that. And this is the law: all cliffs, in some parts of their front, lean backward; and are varied with ledges, bumps, and little shelves or crannies. . . . Provided there be two-inch footholds not more than five or six feet apart going up, or twenty feet apart going down, the Bighorn faces the climb without hesitation. A wide crevice with rugged walls is an easy stairway to this mountaineer. A monkey might use it successfully; I doubt if a cougar could; for he has not the right kind of a mind. A man might have the mental training, but he is wholly without the necessary physical powers.

All this magnificently strenuous physical life is, nevertheless, without much refinement outside the physical realm — partly, perhaps, because the heroic bighorn separates himself during most of the year from "the sex that civilizes ours."

Females go in flocks of from twenty to forty, rams in small bands of five or six. The flock of females is usually led by a bellwether who is probably the grandmother of most of her companions, and in the spring when the young are born the prospective mothers go one by one into retirement to give birth. A few days later the mother will return and introduce the lamb to the flock. But while she is still in seclusion she must perform the odd but indispensable ceremony known to herdsmen of domestic sheep as "owning" the lamb. She must lick it from head to foot, fondle it persistently, and give it to nurse. If for any reason this ceremony is not performed the lamb will die. But most of the mothers' care, other than routine feeding, seems to be concentrated within the brief time during which the "owning" ceremony is performed.

As summer comes on both ram groups and ewe flocks move to a lower altitude, but they do not mingle. By September they are somewhat less aloof from one another, though still keeping to their separate groups. Then in De-

cember, after hard frost and snow, comes the mating season
— which means in this case the season of combat. The ram
is polygamous and like the Homeric hero he collects wives,
more, it seems, as trophies than because he attaches much
importance to any individual for her own sake.

Among most animals which thus fight for wives there is
a good deal of preliminary threatening and boasting. Lions
roar, elks bugle, cattle paw the earth, deer slash and brush
as a challenge. Except perhaps for some stamping and paw-
ing of the ground, the bighorn is a doer not a boaster. Here
is Thompson Seton's account:

> Even in the height of the love-and-war season, the
> ram seems to dispense with the unnecessary embellish-
> ment of loud talk. He looks at his enemy, shakes his
> head, maybe rears for a moment. The other knows
> exactly what that means, and gets ready. They back
> off a little for a good start, and from a hundred feet
> apart, they let loose. With a muzzle velocity of twenty
> miles an hour, they meet like two pile-drivers. The
> crack of horn against horn can be heard two miles off
> on a calm day; each is a 300-pound projectile hurled
> with that fearful force. How can skulls and neck bones
> stand it? Yet they do; and the mighty brutes wheel off,
> ramping on their hind legs, like heraldic unicorns. Each
> tries to show the other fellow how fresh and un-
> winded he is. Then they back up, and go at it again
> — BANG! . . .
> Weight and endurance are what count. They wheel
> and charge again and again, maybe half a dozen times

before they prove the present problem, and show clearly, to the satisfaction of both, that 312 pounds multiplied by twenty miles an hour and backed by ten recuperative kilowatts is better than 340 pounds multiplied by twenty miles an hour and backed by but 5 recuperatives. This is nature's try-out. For this they have trained; and for this, and by this, their race was made, for this they grew those mighty horns. . . .

No one can accuse the "Bélier de Montagne" of being ungallant at this season. He fights to win; and having won, the victor takes the spoils. All of them. Nor are they coy about being taken. For every ounce of strength and unit of desire that is in their strong and amorous make-up is now consecrated to conforming with the scriptural admonition to be fruitful and replenish the earth. The Long-night Moon, December, is the time of fecund passion. A score of wives this grand old Turk may have had: and a score or more young bighorns of his ten-times sifted and selected breed may be born in June, to carry in them the best endowment of the race, so far as he embodied them.

By January, the Snow Moon, the craze is over in the North; and the mighty much-married ram goes calmly back to the club; that is, to his coterie of bachelor friends. He wonders, doubtless, as he passes the ewes, whatever he could have found attractive about those dull creatures not so long ago.

Inevitably what one thinks of is the clash of Homeric chariots or of knights on horseback riding at full tilt to meet in the shock of lance against shield and, possibly, shield against shield. Could they, like the bighorn, have made their armor part of themselves they would doubtless have been glad to do so. The problem in physics which they were determined to solve is of exactly the same sort as that which the bighorn still solves by the same empirical method.

One must not expect a hero to be a courtier or to find the refinements of civilization in a heroic age. Neither Hector nor Achilles would have fit well into a drawing room, and Mark Twain was not the first satirist to remind us that the manners of knights were coarse. So are those of the bighorn. As is the case with most mammals, sexuality expresses itself directly. The bighorn is almost voiceless, and that is just as well, since the sounds made by mammals in time of love are seldom very attractive to us. Like all his fellow mammals except the primates, he is also blind to color and only rank smells have for him any sexual sigficance.

By comparison with the birds he is as uncouth as by comparison with the ant he is unskilled in any of the techniques of agriculture. Furthermore, this primitiveness of his love life and this nonexistence of his technology merely correspond to the primitiveness of his *mores* as a whole. The very fact that he makes few sounds suggests that he has little social feeling despite his loose gregariousness, and there is little sign of social organization. In the ewe flocks the grandmother is followed but can hardly be said to exercise much leadership. Probably in the male groups there is some understood subjection of each individual to the one just above him as there is known to be in the case of the Barbary sheep. But in the bighorn it seems never to have been studied and probably is no more than an analogue of the "peck order" which establishes itself even in a flock of domestic chickens.

In many of these respects the bighorn is even pre-Homeric. He is a hero and a noble savage (or merely "a savage"), depending upon the criteria you set up. Poets can speak for him, but like so many other true heroes he could not make much of a case for himself.

On three different occasions and in connection with as many different topics we have come up against the same question. Are the so-called "higher animals" really higher? And in each case the answer is that by certain objective, easily definable criteria they are not. Some insects have both a far higher social organization and technologically a far more advanced technique for making a living than any other animal except man. Partly because they have also developed astonishingly elaborate methods for assuring the welfare of their progeny they have a better chance of survival than man himself. Hence, by the criteria usually favored by objective scientists, the insects should be called the "highest" animals — highest from the standpoint of the biologist because most likely to survive; highest from the standpoint of the anthropologist because culturally the most advanced. The bighorn may be "noble" but by comparison with the ant he is at most a noble savage. Moreover, while he is disappearing from the earth, the ants are still here in undiminished numbers.

This paradox can hardly be resolved without recourse to metaphysical notions about "value" which the scientist, determined to be "objective," refuses to entertain. Either the "higher" animals are not really higher, or they are higher because they are capable of something we consider more important than either the elaborateness or the "survival value" of the instinctive techniques in which creatures "lower" than they have far surpassed them. If we persist in feeling that they really are "higher" in some meaningful sense it can only be because they seem more human in ways unrelated to complexity of life or even "fitness to survive." Some hint of what that may mean has already been dropped, and if the prejudice in favor of what suggests the human psyche is merely a human prejudice, at least it is inevitable.

Before we consider the question any further it might be amusing to look for the human quality in a vertebrate far below either the bird or the mammal, one in whom the light of the something he shares with us burns very dimly indeed.

7. The Meaning of Awareness

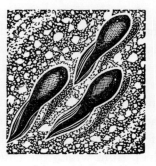

FOR NINE LONG YEARS a large salamander lived her sluggish life in a damp terrarium on my window sill. Before I assumed responsibility for her health and welfare she had lived through a different life — not as different as the life of a butterfly is from that of a caterpillar, but different enough. Once she had lived in water and breathed it. Like her parents before her she still had to keep her skin damp, but now she seldom actually went into the water.

Before she was even an egg her father and her mother, prompted by some no doubt unconscious memory, had left the damp moss or leaves they had normally preferred since achieving maturity and had climbed down into some pond or pool to mate. The prompt result was a cluster of eggs embedded in a mass of jelly much like that which surrounds the eggs of common American frogs. These eggs had hatched into tadpoles easily distinguishable from those destined to become frogs or toads by the two plumes waving from their shoulders — gills for breathing the water which frogs manage to get along without even though they too are temporarily water-breathers.

Most of my specimen's subsequent history was much like that of the young frogs themselves. Legs had budded, and though the tail had not disappeared the plumes had withered away while lungs fit for air-breathing had developed. Sally, as I called her, had then left the water and become a land animal. All this took place quite gradually without any radical dissolution of the organism as a whole, as in the case of the caterpillar, and without the intervention of that dead sleep from which the caterpillar woke to find himself somebody else. Far back in time, Sally's direct ancestors had been the first vertebrates to risk coming to land, and she recapitulated their history.

The rest of my salamander's life was very uneventful but not much more so than it would have been had I left her to her own devices. In fact, returning to the water is almost the only interesting thing the amphibia ever *do*. By comparison with even the butterflies — who lead very uneventful lives as insects go — the amphibia are dull creatures indeed, seemingly without enterprise, aspiration, or any conspicuous resourcefulness.

If you or I had been permitted a brief moment of consciousness sometime about the middle of the Mesozoic era, when the amphibia and the insects were both flourishing, we well might have concluded that the latter were the more promising experiment. I doubt that we would have been very likely to pick out a salamander as our ancestor. Yet the evidence seems pretty definite that nature knew better and that it is from him we come. In Old Testament terms, Amphibia begat Reptile, Reptile begat Mammal, Mammal begat Man.

Even before the Mesozoic was over the beetles were far ahead of the salamanders so far as the techniques of living are concerned. "What," we might well have asked, "do the amphibia have that the insects do not?" What potentiality in them was responsible for the fact that, given the whole Cenozoic still to develop in, the one got no farther than the bee and the ant, while the other has ended — if this is indeed the end — in man?

Perhaps if that anticipatory visit had lasted long enough we could finally have guessed the answer as easily as it can be guessed today by anyone who has kept both insects and salamanders in captivity· and has observed one great difference between them. The insect goes very expertly about his business. But not even those insects who go very expertly about their very complicated business give any sign of awareness of anything not directly connected with that immediate business.

It is not merely that they are absolutely, or almost absolutely, incapable of learning anything. A salamander cannot learn very much either. But the salamander has some awareness of the world outside himself and he has, therefore, the true beginnings of a self — as we understand the term. A butterfly or a beetle does not. Hence you can make a pet out of a salamander — at least to the extent necessary to fulfill the minimum definition of that word. He will come to depend upon you, to profit from your ministrations, and to expect them at appropriate times. An insect is never more than a captive. If you help him he does not know it and he will never come to depend upon your ministrations. He does not even know that you exist. And because of what that implies, a whole great world of experience was opened up to the hierarchy of vertebrates from the salamander on up and has remained closed to the insect.

Seen from the outside, the ants who keep cows, practice agriculture, make war, and capture slaves suggest human beings more strongly than any vertebrate lower than the apes. But if we could see from the inside, the psyche of even the sluggish salamander in my window terrarium would be different. In some dim way she has connected me with herself and I am part of her life.

My old housekeeper used to assure me pridefully from time to time: "She knows me." That, I am afraid, was a bit of overinterpretation. I doubt very seriously that Sally could tell me and my housekeeper apart. But if either one of us approached the terrarium she would rise heavily on her short legs and amble slowly in the direction of the familiar object. We were associated in what little mind she had with the prospect of food.

Was this, many will ask, more than a mere reflex action? Did any such consciousness as I have been assuming really exist? I will not answer that question in the affirmative as positively as many would answer it in the negative. But the consciousness which is so acute in us must have begun dimly somewhere and to me it seems probable that it had already begun at least as far back as the salamanders who lie, though remotely, on our own direct line of descent.

Yet if Sally just barely achieved the status of a pet she fell considerably short of being what we call a domesticated animal. Considerably more awareness of the world around her and considerably more capacity to make an individual adaptation to it would be necessary for that. But because dogs and cats and horses — all of whom have, like us, a salamander in their ancestry — have that considerably greater awareness, they can live a considerable part of our lives and come to seem actual members of our family. Even they are not nearly as ingenious as bees or ants. But we recognize their nearer kinship to us.

Those ants have a culture not only analogous to ours in certain respects but one also far older than ours, since the social insects have been civilized for a much longer time — perhaps thirty times longer — than we have. This was possible because they had settled down biologically — i.e., had ceased to evolve organically — long before we did. And since they were not changing rapidly, they had time to mature and to settle irrevocably into habits and customs, while we are even now still experimenting wildly — discarding habits and techniques every decade or two.

By ant standards we have never had any traditions loyally adhered to. Their so-called virtues — industry, selfless devotion to the good of the community, etc. — are so strikingly superior to ours that certain fanatical critics of human nature and its ways have implied that these creatures whom the biologist calls "lower" are morally "better" than we, and have hoped that in a few million years we might become more like them. Even without going that far and leaving ourselves resolutely out of it as obviously *hors concours,* we may still find ourselves raising again the outrageous question already alluded to. By what right do we call the ants "lower" than, say, a member of a wolf pack? On what basis is the hierarchy established?

Ask that question of a biologist and he will give you ready reasons satisfactory to himself. Anatomically, the insects are simpler. They show very little adaptability. They cannot learn as readily as a wolf can. They can't change their habits very much. They have come to a dead end. They have been precisely what they are for a very long time and will remain that for a very long time to come. "Progress" is something they no longer know anything about. And they are not "intelligent."

All these statements are true enough, but like so many biological distinctions and standards they seem just a little remote. To say that an animal is a compulsory protein feeder is, as we remarked once before, perfectly accurate but has little to do with the rich complex of meanings the word "animal" suggests to the human being who hears it. In certain contexts it is fine. In other contexts — the context of a poem, for instance — it isn't.

Indisputable accuracy does not make it much more satisfactory than Plato's definition of man — a two-legged animal without feathers. Man is certainly that and no other animal is. His definition establishes a criterion that is infallible, but also entirely irrelevant. Apply the test and you will never mistake a wolf or a bird for a man or even mistake a primate — always more or less four-footed — for one of your fellow citizens. This really is a sure way of telling your friends from the apes. But then, you would not be very likely to make a mistake anyway. The definition is perfect but also meaningless.

The explanation a biologist would give why a wolf is "higher" than an ant is almost equally unsatisfactory, because it does not seem to involve the thing on the basis of which we make our judgment. Should they reverse themselves tomorrow and give new reasons in similar terms for deciding that the ant is "higher" we would go right on feeling that he is not. On what, then, is this feeling based if not upon any good scientific criteria? What kind of distinctions appeal to us as genuinely meaningful?

Suppose you play the childish game. Suppose you ask yourself which you would rather be — a farmer ant or a robin. Only the perverse would hesitate. "A robin, of course." But why? What it would come to would certainly be something like this: "Because being a robin would be more fun. Because the robin exhibits the joy of life. Because he seems to be glad to be a robin and because it is hard to believe that an ant is glad to be what he is." Of course we can't say positively that he isn't. We cannot understand his language and he may be proclaiming to the world of other ants with what ecstasy he contemplates the fact that he is one of them. But he cannot communicate with us, and, justifiably or not, we find it hard to believe that he is glad.

Privately, biologists often share our prejudice. But few, I am afraid, would agree to classify animals as "higher" or "lower" on any such basis. They would reply, and rightly so far as biology is concerned, that to say a robin is higher than an ant because he has more joy in living is to cease to be scientific. Also, some might think that it smacks of immoral hedonism. Nevertheless a hierarchy ordered on that basis is meaningful in human terms as the scientific one is not.

If the joy of living is the most enviable good any of the lower animals can attain to and at least the second-best available to man himself, that implies in both a more general capacity which can only be called "awareness" — something that is different from intelligence as usually defined and not perfectly equatable with logic, or insight, or adaptability; also something the salamander has more of than the ant has. There is no way of measuring it, and even the psychologist would be for that reason rather loth to take it much into consideration or even to admit that it exists as distinguished from reason, insight, and the rest. That it does exist in human beings, any contemplative man knows from his own experience.

The best solver of puzzles is not necessarily the man most aware of living. The animal who most skillfully adapts himself to the conditions for survival is not necessarily the one who has the greatest joy in living. And from the standpoint of one kind of interest in living creatures it is perfectly legitimate to think of them as "high" or low" in proportion to the degree of awareness they exhibit.

We can freely admit that the ant's technique of making a living is far more advanced than that of the bird or, indeed, of any vertebrate animal except man. We can see that some species of ants have reached what in terms of human history corresponds to an agricultural society, whereas there is no vertebrate who is not still a mere nomad hunter.

But living — as some men have got around to telling themselves — can be more important than making a living. And making a living seems to be all the ant does, while the robin and many another vertebrate live abundantly.

Yes, I say to myself, the "higher" animals really are higher. Even the sluggish, dim-witted salamander, cold-blooded but vertebrate and with the beginnings of a vertebrate brain, is "higher" than the industrious ant. But it is not for any of the objective reasons either the biologist or the social anthropologist will consent to give that I call him so.

It is because even the salamander has some sort of awareness the insects have not; because, unlike them, he is on his way to intelligence, on his way to pain and pleasure, on his way to courage, and even to a sense of honor as the bighorn is beginning to feel it; on his way to Love, which the birds, bungling parents though they are, can feel and the wise wasp cannot. On the way to the joy of life, which only one or more of these things can make possible.

Once you admit this fact there is something obviously wrong with the orthodox view of the aims and methods of that evolutionary process through which both the blindly efficient ant and the blunderingly emotional bird arrived at their present state. According to that orthodox view "survival value" is the key to everything. But though intelligence does have an obvious survival value, it is by no means obvious that it works any better than the instinct of the insect. As for the emotions, their survival value is not always obvious at all. And if you want to include man in the scheme of evolution, it is so far from obvious that the complexities of civilized emotional and intellectual life have any survival value at all that many recent philosophers have suspected them of being fatal handicaps instead.

This is a fact that raises a question for the evolutionist. If the survival value of intelligence is real enough though no greater than that of instinct, if many of our emotions and the kind of awareness upon which they depend have no obvious survival value at all, then why have certain animals developed both to such a high degree? Why, for that matter, have either they or we developed them at all? Doubtless an intelligent *individual* has a better chance of individual survival than a merely instinctive one. But if nature is careful of the type, careless of the individual, then why should that weigh anything in the scales?

Darwin himself formulated a "law." No organism, he said, ever develops a characteristic beyond the point where it is useful for survival. But, as we have been asking, how useful in that sense is intelligence or even consciousness? Doesn't instinct have an even higher survival value?

It is pretty generally recognized that the insects are the most successful organisms on earth. It is also generally recognized that they get along either with the dimmest consciousness and intelligence, or perhaps without any at all. It is even believed by many that they lost a good deal of what they once had because instinct proved to have a higher survival value. If all this is true does it not suggest that orthodox evolutionism may be in one respect wrong? Does it not suggest that nature (or whatever you want to call it) puts a value on things which do not have any simple survival value? Is it not possible that mammals look after their young with bumbling consciousness rather than with the expertness of instinct because nature has, in some way, been interested not merely in the survival of the fittest, but in "the fittest" for something more than mere survival?

This last question, in a somewhat different form, was actually asked and then left unanswered in the earliest days of Darwinism. Alfred Russel Wallace, generously acknowledged by Darwin as the co-propounder of the theory of natural selection, steadily and from the beginning maintained one difference with his more famous co-worker. It was not and could not be demonstrated, he said, that natural selection could account for "the higher qualities of man." Most notable among these "higher qualities" was, he maintained, the moral sense.

No doubt some manifestations of it had a survival value in society. But not all of them. Man's willingness, sometimes at least, not only to sacrifice himself but to sacrifice himself and others for an ideal, his human conviction that "survival value" is not the only value, did not in themselves have any "survival value." How then could they have arisen if it was, as Darwin said, the inviolable rule of nature that no organism can develop what is not biologically useful to it? An all-inclusive explanation of the phenomenon of life in terms of natural selection would have to account somehow for the very conception of "values which have no survival value." And no such inclusive explanation is forthcoming.

For the most part this question has been simply brushed aside by orthodox evolutionists. Along with other related questions it has been kept alive chiefly by "mere men of letters" — by Samuel Butler, Bergson, Bernard Shaw, and the rest. But it will not down. And there are even signs that some scientists, perhaps especially the neurologists, are less sure than they once were that the mechanistic explanations of all the phenomena of living matter is complete. But if nature has been working toward something besides survival, what is it?

Julian Huxley, one of the most enlightened of present-day evolutionists, has tangled with the question. Evolution, he says, implies progress. But in what does "progress" consist? Certainly, as he admits, it includes something more than a mere progressive increase in the amount of living matter on the earth. That could be achieved by the simplest forms. Nature "wants" not merely more organisms but more complex organisms. But how can it want them if they do not survive more abundantly? Greater complexity implies, he says, "improvement." But what constitutes an "improved" organism? Not, he says, mere complexity itself but a complexity which opens the way to further "improvement." That, it seems, simply closes the circle. The question of what constitutes "improvement" and what sort of values other than mere survival value nature does recognize is still unanswered.

Perhaps the only way to escape from the dilemma that a Huxley recognizes is to make an assumption bolder than he would probably be willing to accept. But the difficulties do vanish if we are willing to accept the possibility that what nature has been working toward is not merely survival; that, ultimately, it is not survival itself but Consciousness and Intelligence *themselves* — partly at least for their own sake.

If Nature has advanced from the inanimate to the animate; if she "prefers" the living to the lifeless and the forms of life which survive rather to those that perish; then there is nothing which forbids the assumption that she also "prefers" conscious intelligence to blind instinct; that just as complex organization was developed even though it had no obvious survival value for the species, so also the awareness of itself which complex organization made possible is also one of her goals.

Whenever man's thinking starts with himself rather than with his possible origins in lower forms of life he usually comes to the conclusion that consciousness is the primary fact. "I think therefore I am" seems to be the most inescapably self-evident of propositions. Only when he starts as far away from himself as possible can he get into the contrary habit of assuming what the nineteenth century did assume: namely, that his own mind is so far from being the most significant thing in the universe that it has no substantial significance at all, being a mere illusion, some sort of insubstantial by-product of those ultimate realities which are unconscious, automatic, and mechanical.

Ever since the seventeenth century, science actually has tended to begin as far away from man himself as possible, while metaphysics has continued to start with man's own mind. Hence the undoubted fact that for a long time, at least, science and metaphysics either grew farther and farther apart, or, as with the positivists, metaphysics simply surrendered to science and tended to become no more than an abstractly stated theory of the validity of science. Yet, as we have just seen, science and positivism leave certain stubborn questions unanswered. Perhaps these questions will ultimately have to be attacked again and from the older point of view.

Aristotle is the acknowledged father of natural history. But because Aristotle lived in an age when it still seemed natural to start with the human mind itself, he reached the conclusion that at least so far as man himself is concerned Contemplation is what he is "for." And if Aristotle had had any clear idea of evolution he would certainly have supposed that a more and more complete awareness, not mere survival, was what nature was aiming at.

Most present-day biologists, following the lead of the nineteenth century, have no patience with any such metaphysical notions. When you come right down to it man is, they say, an animal; and there is only one thing that any animal is "for" — namely, survival and reproduction. Some animals accomplish this purpose in one way and some in another. Man's way happens to involve some consciousness of what he is doing and of why he does it. But that is a mere accident. If what we call intelligence had not had a high survival value it would never have developed. And one of the consequences of this fact is that man is most successful when he uses his intelligence to facilitate his survival. Thinking, or even awareness, for its own sake is a biological mistake. What he is "for" is *doing*, certainly not mooning over what he has done — unless of course that mooning has survival value, as under certain circumstances it may.

What we have been asking is, then, simply this: How good is the evidence — even their own kind of evidence —

which those who take this position can offer in its support? If they are right, then man ought biologically to be the most successful of all animals. No other ought to flourish so exuberantly or have a future which, biologically, looks so bright. But what grounds do we really have for believing anything like that to be the real state of affairs? Does conscious intelligence really work any better than instinct?

No doubt you and I are the most successful of the mammals. When we take possession of any part of this earth the others go into a decline. No bear or wolf, no whale or buffalo, can successfully compete with us. But that doesn't really mean much, because all the mammals are creatures who have already started down the road we have followed so much farther than they. To some considerable extent they too are conscious, intelligent, capable of learning much from experience. Like us they are born with mental slates which, if not entirely blank, have much less written on them than is indelibly inscribed before birth on the nervous systems of many a "lower" animal.

Obviously if you are going to have to depend upon conscious intelligence, then it is an advantage to have that conscious intelligence highly developed. The other mammals over whom we triumph so easily have to fight us chiefly with inferior versions of our own weapons and it is no wonder that they lose. But what of the creatures who learn little or nothing, who can hardly be said to be capable of thought, who are conscious only dimly if at all? Are they really, from the biological standpoint, any less "successful" than we or the other mammals? Can they be said to "succeed" any less well? Are they deprived of anything except consciousness itself?

It is certainly not evident that they are. As a matter of fact the insects are the only conspicuous creatures indubitably holding their own against man. When he matches wits with any of the lower mammals they always lose. But when he matches his wit against the instinct and vitality of the insects he merely holds his own, at best. An individual insect is no match for an individual man. But most species of insects have done very well at holding their own as a species against him. And if you believe the biologists it is only with the prosperity of the species that Nature, or evolution, has ever, or could ever, concern herself.

Who is the more likely to be here on what evolution calls tomorrow — i.e., ten million years hence? Certainly the chance that man will have destroyed himself before then seems greater than the chance that the insects will have done so. Their instincts seem not to have created for them the difficulties and the dangers man's intelligence and emotion have created for him. They have been here much longer than he and it certainly seems not improbable that they will remain here much longer also. As a matter of fact the bacteria are even more "successful" than the insects. There are far more of them alive at this moment than there are of even insects, and it is even more difficult to imagine them ever extinct. If survival is the only thing that counts in nature then why or how did any life higher than that of a bacterium ever come into being?

No answer to that question seems possible unless we are willing to assume that for Nature herself, as well as for us, the instinct of the insect is "better" than the vegetative life of the bacterium, and the conscious concern of the bird for its offspring better than the unconscious efficiency of the wasp. Yet vegetation is not better than instinct and consciousness is not better than instinct if the only criterion is survival value. And if man's mind does not help him to survive more successfully than creatures having no mind at all, then what on earth can it be for? Can it be for anything except itself? Can its value be other than absolute rather than instrumental?

The bird and the man are more successful than the wasp only if you count their consciousness as, itself, some kind of success. The "purpose" of parental concern cannot be merely the successful rearing of offspring, because that can be accomplished quite as successfully without any consciousness at all.

Is it not possible, then, that Aristotle was right, that contemplation is not only the true end of man but the end that has been pursued ever since vertebrates took the road leading to a keener and keener consciousness? Have we been trying to understand the meaning of evolution by beginning at the wrong end? Is it possible that, for instance, the real, the only true "purpose" served by conscious concern over the young is the fact that out of it comes parental love itself? Has what evolution worked toward been not "survival" but "awareness"? Is the ultimate answer to the question "Why is a bungling mammal higher than an efficient wasp" simply that it is higher because it can experience parental love? Was it this, rather than mere survival, that nature was after all along?

8. Undeveloped Potentialities

"THERE ARE many arguments, none of them very good, for having a snake in the house." So Mr. Will Cuppy once wrote, though he was gracious (or is it cynical?) enough to add: "Considering what some do pet, I don't see why they should draw the line at snakes."

Nevertheless one of my friends and neighbors has kept reptiles since he was seven years old and at this moment he has in his house a whole roomful, including some of the largest and most venomous of our native species. I have seen him stopped suddenly during a casual evening walk in the desert by a faint whir; seen him drop to the ground with a flashlight and then rise after a few seconds with a three-foot rattler held firmly and triumphantly just behind the head.

But it is some of his other pets — not of the sort to which Mr. Cuppy rather seems to be alluding — which concern me at the moment. The doorknob on his snake room is placed too high to be reached by his three children, of whom the eldest is now five, but he has reared free in the house two Arizona wildcats which, until they got too big, the children lugged about as though they were unusually phlegmatic tabbies.

Despite the dire prophecies of neighbors, nothing untoward happened to the children; and as for the cats, they went the way of many a domestic Tom. They took to staying out all night; then to staying away for several days at a time; and, finally they did not come home at all. The last time my friend saw one of them he was up a telephone pole about a mile from the house and the other came out of the brush to the call, "Puss, puss."

Their current successor is an eleven-pound, five-month-old male of the same species who is even more completely a member of the family. He is housebroken and most of his behavior is precisely that of a domestic cat. He is very playful and, like many house cats, he walks about the edge of the bathtub while the children are being bathed, apparently fascinated by the human being's strange lack of distaste for water. When he makes a playful leap from five feet away to land on your chest or lap, the impact is considerable; but in many ways he is gentler than a kitten, or, perhaps one should say, seems more aware that his teeth and claws are potentially dangerous. Unlike most kittens, he makes "velvet paws" the inviolable rule and when he kicks with his hind legs in response to being tickled on the belly he keeps the murderous claws of his hind feet carefully sheathed — as a house cat usually does not.

I am not suggesting that always and for everybody a wildcat is a safer pet than *Felis domestica*. It may very well be that the adults are undependable; that they may, as is often said, be capable of sudden flashes of savagery; or, that at the best they are likely to have an inadequate idea of their strength and weight. But to play with so gentle a specimen of what is commonly thought one of the wildest of wild animals is to wonder just how much justification there is for loose talk about "natural ferocity."

What is the "true nature" of this beast? Or is the question as badly phrased as it is when you ask about the "true nature of man"? In both cases the only proper answer may be that both are capable of many different things and that to judge the potentialities of the wildcat by what he is like when truly wild is as misleading as it would be to judge the potentialities of man by studying only the behavior of an Australian bushman.

I have never agreed with those enthusiasts who maintain that a man is "nothing but" what his social environment has made him. But I most certainly do believe that such environment has a good deal to do with his behavior. And though the extent to which the same thing is true of animals is considerably less, the fact remains that you can't know what an animal is really like unless you have known him as different environments have influenced him.

Some scientists are very loth to accept any observations that pet owners have to offer. Not without justification they distrust in fond parents that overinterpretation which is almost as inevitable in the case of a beloved animal as it is in the case of a child. Either is always given the benefit of the doubt. Happy accidents are assumed to be the result of intelligent foresight and the dog lover is often as sure that he can understand the precise semantic significance of a bark as the fond parent of a first child is sure that "aaaa" means "I love Mama."

Despite all this it is still true that the honest observation of well-treated pets can provide data as scientifically valid as any other. What a wild animal is like in the wild can be learned only by observing him in that state; but what you learn is relevant to that state only. In a different environment he becomes something else as surely as a man, moved from one environment to another, changes somewhat as a result of the change.

For the study of anything other than the most mechanically invariable behavior patterns, a pet is certainly a far better subject than a mere laboratory animal. If the pet is, as is sometimes objected, "under artificial conditions," at least these conditions are no more artificial than those of the laboratory and are more likely than the laboratory to expand rather than to freeze latent potentialities. In some ways a pet is more than a wild animal; in a laboratory, any creature is less — as anyone who has ever looked at such a captive with a seeing eye could not have failed to observe. In their loveless imprisonment the more intelligent seem to question and to brood over their Kafka-esque doom. It has imprisoned them for a crime of which they are unaware and prepared for them some future they cannot imagine.

Many of even the animals not commonly made pets respond gladly to the kindlier conditions life with human beings provides. Even in zoos — which are at best little better than model prisons — they develop an awareness of and an interest in people of which we would never suppose them capable if we knew them only in the wild. And it is not by any means merely an interest in the food which human beings sometimes provide. Monkeys are notorious exhibitionists and gibbons, especially, will put on performances at intervals after a preliminary ballyhoo of howls to collect an audience. Less spectacular performances by simpler animals are even more interesting — as, for instance, that regularly indulged in by a group of small ground squirrels in an animal collection with which I happen to be familiar.

These creatures were born in the very glass case where they now live. While still babies they learned to climb up one side of the case, hang momentarily from the roof, and then drop from it to the floor. As time went on they made less and less contact with side or roof until presently they were turning back somersaults with only a kick against the roof at the height of the circle. This performance attracted much attention from the spectators. Now, one need only take up a position in front of the case to have one or more of the squirrels assume his stance and make several rapid flip-overs. Would one ever have suspected from observing these creatures in the wild that they were capable of enjoying the admiration of human beings? Does vainglory find a place in such tiny bosoms? For how many millions of years has the desire to show off existed?

The most objective observer — if he does actually observe — cannot help being struck by the change that comes over an animal who has been really accepted as a companion. Not only cats and dogs but much less likely animals seem to

K

undergo a transformation analogous to that of human beings who are introduced to a more intellectual, more cultivated, more polished society than that in which they grew up.

It is not merely that they adapt themselves to new ways. Their very minds seem to develop and they become more aware of other creatures, including man, as members of a society, not merely as potentially dangerous or useful objects. The human voice, especially, seems to fascinate and delight them. Except, just possibly, for monkeys and dogs, it is doubtful that they ever understand a word as such. The symbolic nature of language seems to be beyond them. But the emotion with which words are spoken communicates directly, and they seem capable of distinguishing many shades of it.

Beyond question it is true at least that they like to be talked to and that talking to them is very important if they are to be drawn into close association with human beings. As I know from experience, this fact is particularly if somewhat unexpectedly conspicuous in the case of geese, though it is much less so in that of ducks. A goose, like a dog, gets a different expression about the eyes if he is accustomed to having some human being speak to him. Thoreau describes the hen wandering about in the kitchen of an Irishman's shanty as "looking too humanized to roast well." "Humanized" is precisely the word, and it is astounding to what extent many animals can become, in that sense, humanized.

There are, of course, limits beyond which no animal can go and there is a great difference between what different species are capable of. The difference between the very slight extent to which my salamander was humanized and what a really intelligent ape is capable of is enormous. But in every case — and this is true of man also — there are limits. To each, nature seems to say: Thus far shalt thou go and no farther. But we never know just how far a man or a beast can go until he has been given a chance. In neither case do the underprivileged furnish a fair answer.

How true this is is well illustrated in the two recent books *King Solomon's Ring* and *Man Meets Dog* by the distinguished Austrian observer Konrad Lorenz. His orthodox training as a scientist warns him against those pitfalls of overinterpretation and anthropomorphism which behaviorists are so eager to expose. But instead of taking it for granted that animal behavior is always explicable in either mechanistic or anthropomorphic terms, he tries to discriminate and to recognize the genuine difference between behavior which is merely mechanically conditioned and that which seems to suggest in the animal rudimentary powers analogous to the human.

Not a man to fear the ridicule of neighbors, he has been observed quacking encouragingly as he crawled through tall grass followed by a line of baby ducks who seemed to take him for their mother. In this instance he was demonstrating not that the ducklings have any evident awareness of the situation, but that they are the victims of what seems to be a merely mechanical conditioning. Like some other animals they are born with a mechanism which is ready to click when they open their eyes. The first moving object they see becomes the thing they are determined to follow. If that first moving object is a man, they will follow him in preference to the mother duck whom they may have seen a few moments later. If, on the other hand, they see her first, it has already become very difficult if not impossible to teach them to follow anything else. This does not mean that all newly born animals have this curious mechanism. Many which have not do nevertheless also operate like a machine, though like a machine differently set up. To account for their behavior or for that of the baby duck it is not necessary to assume any psychic process distinguishable from the mere mechanism of the conditioned reflex.

But Mr. Lorenz does not stop there. He has lived intimately with mature specimens of much more intelligent fowl — especially the highly intelligent raven. Though some of its behavior may be as mechanical as that of the duckling, other actions seem less obviously so. Still others, especially those relating to its intimate association with man and its development of personality, seem to suggest not mechanism but a genuine psychic activity. And it is this which has led Lorenz to invent the new word he uses to describe errors of interpretation which are the opposite of those against which we have so often been warned. "Mechanomorphism," or the stubborn determination to see everything in terms of the machine, may be a fallacy as serious as anthropomorphism.

There are, to be sure, some psychologists who still insist that neither man nor any other animal ever exhibits any behavior not either instinctive or conditioned. To them the final triumph of what they continue to call by the Greek word meaning "the science of the soul" is to have demonstrated that nothing remotely resembling a soul, not even reason or the power of choice, really exists at all. But unless one accepts this thesis the most interesting as well as the least understood branch of psychology is that which does attempt to investigate those actions and those mental states which cannot be shown to be the result of mechanically operating laws.

What makes such studies as those pursued by Lorenz so exceptionally impressive is the fact that he not only recognizes the existence of such phenomena in animals as well as man but also the fact that association with human beings seems to liberate in animals their psychic freedom very much as, so it seems, civilization liberates them in man. Perhaps the chief difference between human and animal psychology, as between the psychology of the lower and the higher animals also, is simply that in both cases what "higher" really means is "exhibiting a more extended range of phenomena which cannot be accounted for in terms of mere 'conditioning.' "

Even more striking in certain respects are two recent English books, Len Howard's *Birds As Individuals*, which describes the surprising and often highly individual behavior of wild birds (especially titmice) who had been left their liberty but invited into the house, and *Sold for a Farthing* by Clare Kipps, which recounts the life story of a tame sparrow. Both these observers are amateurs, but Julian Huxley has vouched for both so far as their reports on the actual behavior of their companions is concerned, and he is ready to grant that they seem to have demonstrated that birds are capable of what might be called "social adjustment" to a degree not hitherto suspected.

What seems to have surprised both Huxley and other commentators even more than the adaptability and apparent intelligence of creatures generally assumed to be less intelligent than many mammals is the extent to which the birds also exhibited individual variation and differences of temperament; the extent, in other words, to which members of the same species seemed to develop what we might as well call "an individual personality." Thus the experience of Miss Howard and Miss Kipps seemed to contradict not only laboratory experiments but also what has been observed of birds in freedom because both suggest that bird behavior is nearly always typical; that individuality hardly exists.

The distinguished American ornithologist Roger Tory Peterson has this to say:

The point that Miss Howard emphasizes . . . is that birds are *individuals*. Their actions often seem to demonstrate some sort of bird intelligence and do not always fall into the oversimplified mechanical patterns which some of us have come to accept. . . . As Dr. Niko Tinbergen, the great behaviorist of Oxford University, comments in *Ibis,* "Miss Howard describes most amazing things, and critical zoologists and psychologists, if not familiar with the ways of birds in the wild, may tend to armchair incredulity. . . . I have no such doubts, however." No other bird book in years has been the subject of so much discussion in England as has Miss Howard's. Some critics may feel that she occasionally resorts to anthropomorphic expressions. . . . Others may differ with her interpretations, but these should in no way be confounded with her facts. Her observations are very careful and her descriptions sensitive and honest. It is a most unusual story she has to tell.

Should we dismiss the new evidence or should we say that the professional observers of wild birds have been wrong?

What puzzled commentators seem to have overlooked is a third possibility. Perhaps Miss Howard's birds not only *seemed* to have more individuality than wild ones but actually did have it. And perhaps they had it simply because they had moved into a social situation where individuality was recognized and given an opportunity to develop. In Bernard Shaw's *Pygmalion* the flower girl who is taught how to pass herself off as a member of sophisticated society announces an important discovery which she has made, namely, that it is not the way you behave but *the way you are treated* which makes the difference between those who are "ladies" and those who are not. Is there any reason why the same should not be, to a lesser extent, true of birds? To reconcile observations made upon any wild animal in a wild environment with those on the same animal after it has become accustomed to treatment as a pet it is not necessary to assume that the original observations were in any way defective or incomplete. Perhaps the seeming conflict merely reveals the fact that animals have a potential capacity for both a degree of individuality and a comprehension of a situation which the circumstances of wild life do not provide an opportunity to develop.

Many modern theorists seem to me to overemphasize a good deal the extent to which men are what "nurture" rather than "nature" makes them. But perhaps in the case of animals we have rather underemphasized it. What it may all come down to is simply that animals, like men, are capable of being *civilized* and that a civilized man or a civilized animal reveals capacities and traits which one would never have suspected in the savage.

We are no longer as surprised as our grandfathers were that an African native who has gone to Oxford can become so typically an Oxford man. Neither by studying an African savage in his native state nor by taking him into a laboratory would we ever be led to suspect the potentialities which can make him an Oxford man. Why then should we find it hard to believe that an analogous change may occur when a wild animal is given an analogous opportunity?

A few years ago I had arranged to spend the night at one of the remote Hopi villages in the house of a young woman educated to be a teacher in a distant Indian school but accustomed to spend her vacations in her native town. When I insisted that for dinner I wanted only that standby of traelers not quite sure of a cuisine, namely ham and eggs, she protested: "After all that is two proteins."

In my instinctive arrogance I had to restrain an impulse to say, "Look here, no Hopi is going to prescribe my diet." But the situation was actually a very minor and superficial example of the phenomena which have compelled us to revise our whole conception of the nature of the difference between civilized and "primitive" minds. So far as difference in mental capacity is concerned it is now generally agreed that there is none — or that the primitives are if anything somewhat our superiors.

If you trust mental tests they seem to demonstrate that the average I.Q. of the Hopi Indian children is higher than that of the white. Most anthropologists seem to agree that there is absolutely no evidence that the human mind, as such, has improved at all during the last five hundred thousand years.

All that civilization has done is to elicit potentialities. Hence there is no reason why a savage, taken young into a sophisticated environment, should not become, as he sometimes does, either very highly civilized or, for that matter, an aesthete and a decadent. Is there, then, any reason for doubting that Miss Howard's birds had become what no one who had observed them only in the wild would have supposed them capable of becoming?

That animals are less plastic than human beings is obvious, and so is the fact that the ultimate development they can reach in the direction of individuality is, absolutely, much more restricted. But their plasticity seems sufficient to make a civilized bird very different from a wild one. And that, of course, is the most striking aspect of the vertebrates' superiority over the superficially more advanced insect. The one has potentialities. The other is fixed and finished. Therefore we might put it this way: some of the limitations of the wild animal are, like those of the savage, merely cultural rather than inherent. Those of the insects are not.

Science recognizes as valid and important the distinction between wild animals and "domesticated" animals. The only other category which it admits, even parenthetically, is that which it labels "pets" and which it dismisses as hardly worthy of scientific consideration. What I am suggesting in all seriousness is that full recogition should be given to that other category which I call "civilized."

The pet may or may not fall within it, depending upon how it is treated. Many show animals, no matter how pampered they may be, are certainly not civilized and indeed hardly deserve even the name of pets. Many a mongrel dog or cat, as well as many another creature of some species only occasionally adopted into human companionship, is much more than a mere pet simply because it is treated with understanding and with love; because it is accepted, if not exactly as an equal, at least with some understanding of the fact that it is capable of responding to a kind of attention and consideration which many kindly people never think of according to even a cherished pet.

The merely domesticated animal, in contrast, is not only something less than a civilized one but something radically different. It has ceased to live the life of nature without being given the opportunity to live any other. Its instincts have faded and its alert senses have sunk into somnolence. Much has been taken away and nothing has been given in return. Such a merely domesticated animal has become a sort of parasite without even developing those special adaptations that make the true parasite interesting in some repulsive way. It is merely parasitic by habit, not by constitution, and of all the animals it is the least rewarding to study because there is almost nothing which can be learned about it. The human analogue is the degenerate remnant of some primitive race which lives upon the fringe of a somewhat more advanced society but has become incapable of leading its own life without having learned any other. Natural history should, on the other hand, no more neglect the civilized animal than anthropology should neglect the civilized man. Without some consideration of both we cannot possibly know what either man or any other animal is capable of.

One of the most important, one of the most fateful developments of thought during the last few centuries has been that which stresses the closer and closer identity of human with animal nature. And that has meant, on the whole, not a greater respect for animal traits and powers and potentialities, but less and less respect for man's. Those potentialities which had once been assumed to be exclusively human now came to be regarded as less and less substantially real. Man was thought of as "nothing but" an animal and the animal was held to be incapable of exhibiting anything except what had formerly been thought of as "our lower nature."

If we are ever to regain a respect for ourselves it may be that we shall regain it by the discovery that the animals themselves exhibit, in rudimentary form, some of the very characteristics and capacities whose existence in ourselves we had come to doubt because we had convinced ourselves that they did not exist in the creatures we assumed to be our ancestors. Even if man is no more than an animal, the animal may be more than we once thought him.

No doubt there are those to whom the concept of the "civilized" animal will seem fantastic and the suggestion that man may regain his self-respect by learning to understand better his animal ancestors even more absurd. To call an animal civilized rather than merely domesticated is, so they will say, to imply that he is to some extent capable of sharing in what are purely human prerogatives. But such an objection is most likely to be raised by the more dogmatic evolutionists, who are, as a matter of fact, those who have the least right to assume a qualitative rather than a merely quantitative difference between the inherent capacities of man and the other animals.

Either man is unique or he isn't. If he is unique then he cannot possibly descend through an unbroken line from the lower animals. If he does descend, or ascend, through such an unbroken line then each of his capacities must have at least its embryonic analogue in the simpler creatures who preceded him.

Evolution implies development, not the sudden appearance of something totally new. No such totally new capacity could have evolved at all, but would have had to have come suddenly into being. And one of the most important of man's capacities is that which enables him to do more than simply "adapt himself to changed conditions." It is the capacity to develop those unsuspected potentialities that make "civilized" mean something more than merely "adapted to group life in a technologically complex society." An essential part of that something more is the development of a more varied, more vivid psychic life. And this is precisely the something more which the civilized animal also unexpectedly manifests.

9. Reverence for Life

THE VANDAL
AND THE SPORTSMAN

IT WOULD NOT BE quite true to say that "some of my best friends are hunters." Nevertheless, I do number among my respected acquaintances some who not only kill for the sake of killing but count it among their keenest pleasures. I can think of no better illustration of the fact that men may be separated at some point by a fathomless abyss yet share elsewhere much common ground.

To me it is inconceivable how anyone should think an animal more interesting dead than alive. I can also easily prove to my own satisfaction that killing "for sport" is the perfect type of that pure evil for which metaphysicians have sometimes sought.

Most wicked deeds are done because the doer proposes some good to himself. The liar lies to gain some end; the swindler and the thief want things which, if honestly got, might be good in themselves. Even the murderer may be removing an impediment to normal desires or gaining possession of something which his victim keeps from him. None of these usually does evil for evil's sake. They are selfish or unscrupulous, but their deeds are not gratuitously evil. The killer for sport has no such comprehensible motive. He prefers death to life, darkness to light. He gets nothing except the satisfaction of saying, "Something which wanted to live is dead. There is that much less vitality, consciousness, and, perhaps, joy in the universe. I am the Spirit that Denies." When a man wantonly destroys one of the works of man we call him Vandal. When he wantonly destroys one of the works of God we call him Sportsman.

The hunter-for-food may be as wicked and as misguided as vegetarians sometimes say; but he does not kill for the sake of killing. The rancher and the farmer who exterminate all living things not immediately profitable to them may sometimes be working against their own best interests; but whether they are or are not they hope to achieve some supposed good by their exterminations. If to do evil not in the hope of gain but for evil's sake involves the deepest guilt by which man can be stained, then killing for killing's sake is a terrifying phenomenon and as strong a proof as we could have of that "reality of evil" with which present-day theologians are again concerned.

Despite all this I know that sportsmen are not necessarily monsters. Even if the logic of my position is unassailable, the fact still remains that men are not logical creatures; that most if not all are blind to much they might be expected to see and are habitually inconsistent; that both the blind spots and the inconsistencies vary from person to person.

To say as we all do: "Any man who would do A would do B" is to state a proposition mercifully proved false almost

as often as it is stated. The murderer is not necessarily a
liar any more than the liar is necessarily a murderer, and
few men feel that if they break one commandment there
is little use in keeping the others. Many have been known
to say that they considered adultery worse than homicide
but not all adulterers are potential murderers and there are
even murderers to whom incontinence would be unthinkable. So the sportsman may exhibit any of the virtues —
including compassion and respect for life — everywhere except in connection with his "sporting" activities. It may
even be often enough true that, as "antisentimentalists" are
fond of pointing out, those tenderest toward animals are
not necessarily most philanthropic. They no more than
sportsmen are always consistent.

When the Winchester gun company makes a propaganda
movie concluding with a scene in which a "typical American
boy" shoots a number of quail and when it then ends with
the slogan "Go hunting with your boy and you'll never have
to go hunting for him," I may suspect that the gun company
is moved by a desire to sell more guns at least as much as by
a determination to do what it can toward reducing the in-
cidence of delinquency. I will certainly add also my belief
that there are even better ways of diminishing the likelihood
that a boy will grow up to do even worse things. Though it
seems to me that he is being taught a pure evil I know that
he will not necessarily cultivate a taste for all or, for that
matter, any one of the innumerable other forms under which
evil may be loved.

There is no doubt that contemporary civilization finds a
place in the vaguely formulated code of ethics to which

L

the majority gives at least formal assent for what it is most likely to call "kindness to animals." It is equally obvious that this degree of recognition is very recent. The Old Testament does say that the virtuous man is kind to his beasts. Thomas Aquinas did disapprove of cruelty to animals, though only on the ground that it would lead to cruelty to men. But there is very little evidence in Western culture before the eighteenth century that the torture (much less the killing) of animals for sport ordinarily revolted anyone. English law, though it was lagging as usual somewhat behind enlightened opinion, did not forbid even the most sadistic abuse of animals for pure pleasure until 1822 when Parliament passed the first law protecting any animal against cruelty per se. Even today when the grosser, more gratuitous forms of needlessly inflicted pain are generally recognized as evil, the difference in attitudes toward killing for sport range through a gamut that must be as wide as it ever was.

Examples of three different but typical ways of refusing to acknowledge that any defense of such killing is called for may be plucked out of recent popular periodicals.

In the spring of 1955 a magazine called *Sports Illustrated* distributed a questionnaire intended to determine the public attitude toward hunting. An answer received from a woman in Tampa, Florida, was as follows: "I am not the sloppy, sentimental type that thinks it's terrible to shoot birds or animals. What else are they good for?" And *The New Yorker,* which reprinted her reply, answered the question with an irony likely to be lost on the asker: "Bulls can be baited by fierce dogs, and horses sometimes pay money."

About a year before, *The New Yorker* had also, though without comment and merely in the course of a report on the personality of the new United Kingdom's Permanent Representative to the United Nations, quoted Sir Pierson Dixon as remarking genially, apropos of some articles on sport which he had written for English periodicals: "I like this shooting thing, stalking some relatively large animal or, even more enjoyable, shooting birds. It's like the pleasure of hitting a ball."

A little later *Time* Magazine ran an article about how duck hunters near Utah's Bear River Migratory Bird Refuge (*sic*) "could hardly shoot fast enough" to bring down the ducks they found there and it adorned the article with a quotation from Ernest Hemingway's "Fathers and Sons": "When you have shot one bird flying you have shot all birds flying, they are all different and they fly different ways but the sensation is the same and the last one is as good as the first."

Of these three attitudes the first may seem the simplest and the most elementary, but perhaps it is not. The blank assumption that the universe has no conceivable use or meaning except in relation to man may be instinctive; nevertheless, the lady from Tampa is speaking not merely from naïveté. She is also speaking for all those minds still tinctured by the thought of the medieval philosophers who consciously undertook to explain in detail the *raison d'être* of the curious world of nature by asking for what human use God had created each species of plant or animal. If any given creature seems good for nothing except "sport" then it must be for sport that it was created.

Hemingway's utterance, on the other hand, is the most sophisticated of the three and the only one that seems to make the pure pleasure of killing a consciously recognized factor. The mental processes of the Permanent Representative are neither so corrupt as those of Mr. Hemingway nor so intellectually complicated as those of the lady from Tampa. He is not, like the first, looking for madder music and stronger wine, nor, like the second, attempting to answer the philosophical question of what animals and birds "are for." Because of the dreadful uncomprehending innocence sometimes said to be found most frequently in the English gentleman it has simply never occurred to him that the creatures whom he pursues are alive at all — as his phrase "like the pleasure of hitting a ball" reveals. Birds are simply livelier, less predictable clay pigeons. And it is in exactly the same light that those of his class have sometimes regarded the lesser breeds without the law, or even the nearly inanimate members of all the social classes below them.

For the attitude farthest removed from this, Albert Schweitzer is the best-known contemporary spokesman. But one can hardly have "reverence for life" without some vivid sense that life exists even in "the lower animals" and it is this vivid sense that is lacking in the vast majority of sportsmen and equally in, say, the abandoners of pets and, not infrequently, one kind of biological scientist. Often not one of them is so much as tinged with the sadism which Hemingway's opinions and activities seem to suggest. It is not that they do not care what the abandoned pet or the experimental animal suffers but that they do not really believe he suffers to any considerable degree. In the case of the hunter it is often not so much that he wants to kill as that he has no vivid sense that he is killing. For him, as for Sir Pierson, it is more or less like "hitting a ball."

The conviction that "man is an animal" is certainly very widely held today, and the further conviction that man is "nothing but an animal" only somewhat less so. One might expect that these convictions would lead to the feeling that no benevolence, philanthropy, humanitarianism, "good will," or even simple decency can logically stop short at the line which separates the highest animal from those next below him and that less scrupulosity in dealing with any creature is justifiable only in proportion to that creature's lesser degree of sensibility and aliveness. Between the human being assumed to be only a species of animal and any other species no qualitative rather than merely quantitative distinction can logically be assumed to exist, and the wanton killing of an animal differs from the wanton killing of a human being only in degree. The two murders are not equally wicked but they are not wholly different either. Such wanton killing of an animal out of no necessity and not even for the sake of a minor need is at least a small murder — not an innocent game.

If the popularization of the biological sciences has not done as much as might be expected to foster some such convictions, that is in part the result of the attitude these sciences themselves have often taken. One of the intentions seldom absent from the pages of this book has been the intention to suggest in as many different ways as possible that to call man "an animal" is to endow him with a heritage so rich that his potentialities seem hardly less than when he was called the son of God. Much biological science has on the contrary tended to draw diametrically opposite conclusions, and not only to deny man the divine origin once assumed to be his but to deprive the animal kingdom to which he is assumed to belong of the powers which, during most of human history, it had been assumed to share in some degree with humanity.

Cartesianism maintained that every animal other than man was a machine, though man was not. But when Cartesianism was rejected by science, science took more away from man than it restored to the lower animals. Increasingly, at least until very recent years, it minimized and often denied the effectiveness of reason or the significance of consciousness in man himself and not surprisingly denounced as absurd the attribution to animals of what it hardly admitted as real in man. We were urged to study both him and the other animals exclusively in terms of instincts, conditioned reflexes, and behavior patterns. Hence it is not surprising that the ultimate result was not the treatment of animals with some of the tenderness and consideration due to men but, in the totalitarian states where science was for the first time freed from all religious, philosophical, or merely literary restraints, to treat men with the brutality usually adopted toward animals.

When Thoreau allowed himself to be persuaded to send a turtle as a specimen to the zoologists at Harvard he felt that he had "a murderer's experience in a degree" and that however his specimen might serve science he himself and his relation to nature would be the worse for what he had done. "I pray," he wrote, "that I may walk more innocently and serenely through nature. No reasoning whatever reconciles me to this act."

In general, however, professional students of living things are only somewhat more likely than the average man to feel strongly any "reverence for life." One of the most distinguished American students of birds told me that he saw no incompatibility whatever between his interest in birds and his love of "sport." Many, perhaps most professional students find no reason too trivial to "collect" a bird or animal, though their habitual use of this weasel word may suggest a defensive attitude. And I have often wondered that sportsmen who find themselves subject to many restrictions have not protested as unfair the "collector's licenses" rather freely granted and sometimes permitting the holder to shoot almost anything almost anywhere and at any time.

Audubon slaughtered birds in wholesale lots so large he once remarked that birds must be very scarce when he did not shoot more than a hundred a day. Even the more exuberant of today's collectors are more aware that species are exhaustible, but many private collectors, as well as museums, are proud of their trays containing thousands of bird skins. Whatever may be said in justification of science's need for study specimens, it is still worth noting that the more trivial the question being studied the more lives it is likely to cost — as when, for example, an earnest student shoots hundreds of birds or animals just to find out whether or not he can establish a regional variety by proving that the average size of some item in the pattern of plumage or

pelt is a millimeter greater in one geographical area than
in another. Moreover, the record comes to seem to many
more important than the living thing recorded and it is an
accepted principle among ornithologists that since "sight
records" can always be disputed it is the bounden duty of
every student to shoot immediately any species believed to
be hitherto unreported in the region where it is observed.

The first new species known to have settled without human aid in the United States is the cattle egret first seen in
Massachusetts in April, 1952 — at which time it was duti-
fully shot by the first knowledgeable bird-lover to see it.
Such was his way of saying "Welcome Stranger." By now
the species is said to be well established in Florida. Was the
desire to demonstrate the identity of the first observed specimen worth the further hardening of the heart it cost? Thoreau would have thought not. What little science gained in
this particular instance was just possibly not worth what a
bird lost.

Obviously the problem raised by all this is not solvable in
any clear-cut way. The degree of "reverence for life" which
man or any other animal can exhibit is limited by the facts
of a world he never made. When it was said that the lion
and the lamb shall lie down together, the hope that they may
someday do so carries with it the obvious implication that
they cannot do so now. Even Albert Schweitzer's rule that
no life shall be destroyed except in the service of some
higher life will be differently interpreted almost from individual to individual.

Just how great must be the good that will accrue to the

higher animal? Interpreted as strictly as possible, his law would permit killing only in the face of the most desperate and immediate necessity. Interpreted loosely enough, it might justify the slaughter of the twenty thousand birds of paradise, the forty thousand hummingbirds, and the thirty thousand birds of other species said to have been killed to supply the London feather market alone in the single year 1914. After all, even fashionable ladies are presumably "higher" than birds and they presumably took keen delight in the adornments which the birds were sacrificed to provide.

Some pragmatic solution of the rights of man versus the rights of other living creatures does nevertheless have to be made. Undoubtedly it changes from time to time and it is well that the existing solution should be re-examined periodically. Because the 1914 solution was re-examined, comparatively few birds are killed for their feathers and it is not demonstrable that the female population is any the worse for the fact.

In India members of the Jain sect sometimes live on liquid food sipped through a veil in order to avoid the possibility that they might inadvertently swallow a gnat. There are always "antisentimentalists" who protest against any cultivation of scruples on the ground that they can logically lead only to some such preposterous scrupulosity. But there are extremes at both ends. Those who have scruples are no more likely to end as Jains than those who reject all scruples are likely to end as Adolf Hitlers. The only possible absolutes are reverence for all life and contempt for all life and of these the first is certainly no more to be feared than the second. If there is any such thing as a wise compromise it is not likely to be reached by the refusal to think. However difficult it may be to draw lines, they have to be drawn and draw them most men do, either thoughtlessly or with thought.

One may conclude that mankind could not continue to exist without killing some living creatures, just as one may conclude that without a willingness to resort to war self-preservation for a nation is impossible. Killing is evil and war is hell. But it makes a tremendous difference whether or not one concludes from these facts that any concern with honor and mercy or any refusal to accept any methods or countenance all cruelty is foolish. It makes the same sort of difference whether you say that life in the world is impossible without some self-regarding worldliness and that therefore men can never be other than remorseless egotists anywhere, or whether you say that, despite the cruel dilemmas with which a contingent universe continually confronts us, we can still sometimes elude or mitigate them. Though the lion and the lamb may never lie down together it may still be that the more we do elude or mitigate the implications of that fact the better it is for man and beast alike.

In his actual practice, civilized man has been more ruthlessly wasteful and grasping in his attitude toward the natural world than has served even his most material best interests. The more we learn about the interdependence of living things, the clearer it becomes that the practical utility of the land upon which we live has been diminished seriously by the determination to allow it to serve no purposes but our own. Many once prosperous lands have, like Spain, found poverty and famine stalking across their fields and pastures without knowing what had brought the calamity upon them. In the United States many ranchers and many farmers have been invoking the same specter despite full knowledge of what they were doing and because they were determined merely to get theirs while the getting is good.

Possibly — as some hope — a mere enlightened selfishness will save them in time. Perhaps it will teach them to save their soil from direct exhaustion and furthermore leave uncut upon the mountains some of the trees without which the rushing torrents from mountain rains will ultimately wash their farms away. Possibly, even, it will teach them also that the fox or the coyote who occasionally eats one of their hens or their sheep eats more often the rodents they will have to struggle less successfully against when they have eliminated all the predators. In Deuteronomy the husbandman is forbidden to glean his cornfields where the scattered grain belongs by right to the poor. Surely, even when no direct utility is obvious, it is a sin to grudge the small fellow creatures our crumbs and grudge even the wild flower its few inches beside the farmer's fence or along our roadside.

If the earth is still livable and in many places still beautiful, that is chiefly because man's power to lay it waste has been limited. Up until now nature has been too large, too abundant and too resistant to be conquered. As Havelock Ellis once wrote without exaggeration, "The sun, moon and stars would have disappeared long ago if they had happened to be within reach of predatory human hands."

Such predatory human hands have exterminated many kinds of living creatures and rendered many a flourishing acre barren — but not so many as they would have destroyed had the reach of the hands not been limited. Our numbers and our ingenuity have been growing at a prodigious rate. We may not have progressed as far in certain directions as we commonly suppose, but there is no doubt about the reality of Progress so far as the power to destroy is concerned. The day is fast approaching when nature's resilience will no longer protect us from ourselves. We are on the point of being able actually to do what for several centuries we have dreamed of doing — namely, "conquer nature." And we may be reminded too late that "to conquer" means to have at least the ability to destroy. If the mass of men continue to be what they have long been that ability will be used.

Even if we should learn just in the nick of time not to destroy what is necessary for our own preservation, the mere determination to survive is not sufficient to save very much of the variety and the beauty of the natural world. They can be preserved only if man feels the necessity of sharing the earth with at least some of his fellow creatures to be a privilege rather than an irritation. And he is not likely to feel that without something more than the intellectual curiosity which is itself far from universal today. That something more you may call Love, fellow-feeling, or "reverence for life," though — as was recently pointed out in a letter to a philological journal — Schweitzer's own term *Ehrfurcht* carries a stronger sense of "awe" than the English word that has been weakened in use.

And whatever you call it, it is something against which both urban life and some of the intellectual tendencies of our times tend to militate. Increasing awareness of what the science of ecology teaches promises to have some effect upon the public's understanding of the practical necessity of paying some attention to the balance of nature. But without reverence or love it can come to be no more than a shrewder exploitation of what it would be better to admire, to enjoy, and to share in.

Unfortunately the scientific study of living creatures does not always promote either reverence or love, even when it is not wholly utilitarian in its emphasis. It was the seventeenth-century naturalist John Ray who first gave wide currency in England to the conviction that God made other living things not exclusively for the use of man but also for both his delight and for theirs.

Unfortunately that laboratory biology which has tended to become the most earnestly cultivated kind of scientific study is precisely the kind least likely to stimulate compassion, love, or reverence for the creatures it studies. Those who interested themselves in old-fashioned natural history were brought into intimate association with animals and plants. Its aims and its methods demanded an awareness of the living thing *as* a living thing and, at least until the rise of behaviorism, the suffering and the joy of the lesser creatures was a part of the naturalist's subject matter. But the laboratory scientist is not of necessity drawn into any emotional relationship with animals or plants and the experiments which of necessity he must perform are more likely to make him more rather than less callous than the ordinary man.

At best, compassion, reverence for life, and a sense of the community of living things are not an essential part of his business as they are of the more vaguely defined discipline of the naturalist. And for that reason it is a great pity that the most humane and liberal of the natural sciences should play so small a role in the liberal arts curriculum. While still under the influence of an older tradition, field botany and field zoology were quite commonly taught in American colleges even in the remoter parts of the United States. Today few liberal arts undergraduates know anything of such subjects and often would find no courses open to them if they did.

In Columbia College, with which I happen to be most familiar, there are elementary courses available in biology, but they are designed primarily to meet the needs of those headed for the medical and other professional schools. They introduce the student to the subject via the anatomy and physiology of the lower animals and are confined so closely to dissection and other laboratory operations that he is seldom if ever brought into contact with a living plant or an animal living a natural life. Undoubtedly these courses serve their purpose, but they have little to contribute to education in the humanities and those committed to such an education very rarely attend them. If such a student should feel, for example, that he ought not be as ignorant as most of his fellows concerning the plants, animals, and birds which figure so largely in both British and American literature, he would get little help from the curriculum. There are many courses in "The Nature Poets" in American colleges. But nature is usually left out of them.

Very recently I had occasion to spend a week on the campus of one of the oldest and most respected of the smaller liberal arts colleges of the eastern seaboard. It was one that prides itself on its exclusive concern with liberal rather than preprofessional education. A benefactor gave it some years ago a beautiful wooded tract adjoining the campus which is lavishly planted with native and exotic flowering trees and shrubs. When no student or teacher with whom I had been brought into contact could tell me the name of an especially striking tree, I sought out the head of the botany department, who was also its only member.

He smiled rather complacently and gave this reply to my question: "Haven't the least idea. I am a cytologist and I don't suppose I could recognize a dozen plants by sight." The secrets of the cell are a vastly complicated and important subject. But should they be the one and only thing connected with plant life which a student seeking a liberal education is given the opportunity to learn?

That a similar situation does not always prevail I know from observation, but when it does not that is usually simply because the teacher employed happens to have a broader interest, not because those in charge of the curriculum are convinced that some knowledge of the natural world is a part of a liberal education. Biology as commonly taught is not a humane subject; it is simply an elementary preparation for the trade of the specialist.

To proceed from the dissection of earthworms to the dissection of cats — both supplied to hundreds of schools and colleges by the large biological supply houses — is not necessarily to learn reverence for life or to develop any of the various kinds of "feeling for nature" which many of the old naturalists believed to be the essential thing. To expect such courses to do anything of the sort is as sensible as it would be to expect an apprenticed embalmer to emerge with a greater love and respect for his fellow man. And an increased love or respect for living creatures is one of the last things many college courses in biology would propose to themselves.

"Nature study" is often relegated to the lower levels and sometimes thought of as being really appropriate only to the kindergarten. Even in the elementary grades the tendency to devote more attention to dead animals than to living ones sometimes makes its appearance. In a very "progressive" school I have seen teen-agers introduced to the old dreary business of dissecting earthworms; and there are worse things than that when bungling, pointless experiments upon living animals are encouraged. The catalogue of a leading biological supply house boasts of the wide increase in the use of "nutrition experiments" (grandly so-called) in schools. It offers eight different deficiency diets together with the living animals whose malnutrition, when they are fed any one of these diets, may be observed by the curious. Very recently the head of the National Cancer Institute urged high school teachers to teach their pupils how to produce cancer in mice by the transplantation of tumors and in chicks by the injection of enzymes.

Is it sentimental to ask whether anyone not preparing for the serious study of anatomy is likely to be any the better for the dissecting of a cat, or whether anyone, no matter what career he may be preparing himself for, is any the better for having starved a rat or induced cancer in a mouse? However completely experiments up to and including vivisection may have justified themselves, is there any possible excuse for repeating them merely by way of spectacle?

By now it is as well known that a rat will sicken and die without certain minerals and vitamins as it is that he will die if given no food at all. Would anyone learn anything by poking out eyes in order to prove that without them animals can't see? Or, for that matter, from undertaking to find out for himself whether or not it is really true that even Jews can bleed? Yet to deprive animals of protein is hardly more instructive. Taught by such methods, biology not only fails to promote reverence for life but encourages the tendency to blaspheme it. Instead of increasing empathy it destroys it. Instead of enlarging our sympathy it hardens the heart.

The grand question remains whether most people actually *want* hearts to be tenderer or harder. Do we want a civilization that will move toward some more intimate relation with the natural world, or do we want one that will continue to detach and isolate itself from both a dependence upon and a sympathy with that community of which we were originally a part? Do we want a physical environment more and more exclusively man-made and an intellectual, emotional, and aesthetic life which has renounced as completely as possible its interest in everything inherited from the long centuries during which we were, willy-nilly, dependent upon what the natural world supplied? Do we want cities

M

completely sterilized and mechanized; do we want art that imitates exclusively the man-made rather than the natural?

There is a sizable minority which has asked itself these questions and answered them with an unqualified "Yes." There is another minority, perhaps almost as large, which answers them with an equally definite "No." But the large majority has never faced these questions in any general form, though it is nearly everywhere drifting without protest toward a pragmatic affirmative.

No doubt all societies not completely static exhibit many of what the Marxists call "inner contradictions." In our own, some of the less spectacular of these contradictions involve the questions just asked. We set aside wild areas and then "improve" them out of all wildness. We teach kindness to animals and even reverence for life, but we also believe that fathers should teach their sons to hunt or encourage them to dissect cats and watch rats starve. Moreover, these contradictions flourish even within organizations that seem at first sight ranged on one side or another, and the organizations are often supported by an uneasy united front composed of what are really antithetical parties.

What, for example, are the national and the various state conservation and wildlife departments for? Are they to preserve wildlife or to provide game for hunters to kill? If for the latter, then is the justification the beneficial effects of sport or is it the contribution to the general economic prosperity made by the arms industry? When there is a conflict, what comes first?

It would be difficult to get from many organizations a clear-cut statement and I have been told of at least one instance where an officer of a state commission protested hotly against the exhibition in a state park of a young deer which children were allowed to pet because, so he said, making pets of wild animals creates a prejudice against hunting. And to leave no doubt concerning the ultimate reason for his attitude he is said to have added, "After all, guns and ammunition are big business."

I have no doubt that many of his colleagues would violently disagree but I have also been told (though this is for me mere hearsay) that when Walt Disney issued the early animal film *Bambi* he received *protests* from many quarters. Somebody can make more money out of slaughtering animals than anyone can make out of loving them. The hard heart is more economically productive than the tender one. The sporting instinct pays off. Reverence for life does not.

Upon whatever basis all the inner contradictions are ultimately resolved, the consequences for the physical, intellectual, and emotional life of tomorrow's man will not be trivial. The sum of all the resolutions will help determine whether we have decided to go it completely alone and to depend no longer upon nature for food, health, joy, or beauty.

An obviously unfriendly reporter revealed not long ago that President Eisenhower had ordered removed from the White House lawn the squirrels which were interfering with his putting green, and even so trivial an incident is a straw in the wind. To hold golf obviously more important than squirrels indicates a tiny but significant decision. It points toward a coming world where there will be more golf courses and fewer wild plants as well as wild animals — hence to a world less interesting and less rich for those who would rather hunt a flower or watch the scamperings of a squirrel than chivy a rubber ball over a close-cropped grass plot.

The late David Fairchild, who was responsible for the introduction of so many useful and beautiful plants into the United States, tells the story of an army officer assigned to an office building in Miami during the First World War.

"I haven't got anything but human beings around me in that building where I spend my days. Aside from the floor and the ceiling, the doors and windows and desk and some chairs there isn't anything but people. The other evening when I was feeling particularly fed up with the monotony of the place, I went into the lavatory and as I was washing my hands a cockroach ran up the wall. 'Thank God for a cockroach!' I said to myself, 'I'm glad there is something alive besides human beings in this building.'"

It may well be with such small consolations that the nature-lover of the not too distant future will be compelled to content himself. Cockroaches will not easily be exterminated.

10. Devolution

ON THE SPRING MORNING when I began writing this book I
might have picked the illustrations for most of my themes
within two hundred yards of my window. Just about that
far away several colonies of ants were practicing their incredible agriculture, and even closer at hand a jack rabbit —
who had casually given birth to her young on the bare
ground — was exemplifying the socially retarded mammal.
A tiny lily pool was teeming with protozoa, and it is quite
possible, though I did not look at the moment, that among
them were some specimens of Volvox demonstrating how
death and sex had been invented. Just below the surface of
the ground the large green caterpillars who had grown fat
the summer before on the poisonous leaves of a Sacred
Datura which I had transplanted for the sake of its huge
white trumpet flowers, were sleeping out the big sleep from
which they would awake presently looking like somebody

else and proving that in nature the child is not always father to the man.

Because I have no lawn I had no dandelions, but my nearest neighbor, who lives not very far away across the desert, does have one. Therefore, of course, and despite all her efforts, she has dandelions too. And thereby hangs another tale.

This tale is, for a change, a sinister one. Nothing looks more innocent than a dandelion, but instead of pointing back to an unimaginably remote past as Volvox does dandelions may be pointing forward to a remote possible future. And that future has its disturbing aspects. If Volvox seems to hold the promise of human beings, human traits, and even human values, the dandelion seems to threaten them all.

According to those who ought to know, it is one of the "highest" of all plants. That means, first, that its life mechanisms are among those which evolution has most recently developed; and, second, that all these recent devices have an extraordinarily high survival value. As a race, dandelions are prospering and inheriting the earth. In fact they are terrifyingly efficient. They suggest something which life may be coming to, instead of, like Volvox, what it came from. But they have achieved their efficiency by casting aside certain of the characteristics, tendencies, and traits that embody what human beings regard as valuable and are prone to think of as "higher" than the mere efficiency of this "high" plant.

When Volvox, or something more or less like it, invented sex it took a tremendous step. Biologically, sexuality proved so useful that for many millions of years it accompanied and helped make possible every biological advance. And because Volvox, or rather its lost co-inventor of sex, was neither definitely plant nor animal but somewhere in the family tree of plants and animals alike, it could set both plants and animals to developing along the lines sexuality made possible. The egg and the sperm are standard equipment throughout most of the vegetable as well as most of the animal world. Love was the mother of all things — or at least of all living things above the level of the very simplest.

Moreover, its psychic consequences were unimaginably portentous. Without sexual love there might never have been love of any other kind. Without it we cannot imagine what either civilization or the inner life of man would have become. It was the sole cause of beauty in flowers, as well perhaps as of the most important part of what men call poetry. *But love is not the mother of dandelions.* And that is the first of a whole series of sinister things which can be said about them.

Love was the mother of the dandelions' ancestors. It enabled them to become what they now are. But they found that they could get along without it. They turned their backs upon the whole history of living things as it had unrolled since the first days of Volvox. They proved that Sex is *not* necessary. And they suggest the possibility that in some future no remoter than the past we look back upon it may disappear again — with consequences as far-reaching as those which followed its introduction.

That dandelions really are sexless you would never guess from a superficial acquaintance. They bear a multitude of brightly colored flowers and in most plants what we call flowers are, biologically, merely sex organs which have clothed themselves in gay colors and bathed themselves in pleasant odors — having done both in order to attract the insects or birds (even in some cases mammals), which proved to be useful in assuring the pollinization upon which fertility depends.

Before flowers were invented most plants had depended upon the wind to perform this function in a wasteful, haphazard fashion that compelled them to squander large quantities of pollen. The modern representatives of some ancient plants — notably the conifers — still produce nothing which we recognize as a flower. But once insect pollination was discovered, the flowers went on during millions of years, becoming more and more elaborate. During all this time biological usefulness, or efficiency, went hand in hand with beauty. Everything we still find beautiful in flowers is a direct consequence of this fact. We sometimes say that among animals sexuality "flowered" into Love; among plants it flowered literally into flowers. But in the dandelion Evolution has turned about and headed back toward the past again.

Suppose you pluck one of the dandelion's so-called flowers and tear it apart. Then you will see that what you thought was a flower is actually a close-packed head of many separate flowers. Attached to each not-yet-matured seed will be one of those yellow blades you would naturally call a petal. It turns out to be not a petal in the usual sense but a fused group of five petals, whose former existence as separate entities is now revealed only by the little notches which persist along the upper edge of the blade.

Less conspicuous but far more important to the dandelion is the fact that each immature seed also bears at its

summit a circlet of developing plumes which correspond anatomically to the green cup called a "calyx" in more conven-
tional flowers. They will become in time the parachutes that
will scatter the seeds literally to the four winds. And here
again the dandelion is unusually efficient. Seeds of any kind
were invented only after plants had experimented for millions of years with other methods of sexual reproduction and
seeds had been used for millions of years more before any
plant developed any device to assure their distribution as advanced technologically as the dandelion's parachute.

What all this adds up to is the fact that our dandelion belongs to the great modern plant family called "composite,"
whose unifying characteristic is the fact that what looks like
their flower is really a head of simplified flowerets. Theirs
is a very efficient method of seed production. But in many
cases the petals are merely vestigiary as they are in the dandelion, to whom it makes not the slightest difference whether
or not it is ever visited by an insect; and, in the case of some
other members of the composite family, petals have disappeared altogether.

Why they can thus be dispensed with will be clear very
soon. For the moment the important thing is that in this
most recent, most up-to-date of all plant groups, the flower
is, for all its efficiency, devolving rather than evolving —
aesthetically at least. Throughout a tremendous stretch of
time beauty and efficiency had gone hand in hand. Here
the trend has been reversed — so definitely and with such
a gain in efficiency that flowers may be on the way out. The
earth may someday again be flowerless as for many millions
of years it once was.

This aesthetic devolution was made possible by, and proceeded step by step just behind, a decreasing emphasis on sexuality. Even wind pollination had been a step forward in the process that began when separate male and female sex organs were developed. It increased the probability that the egg destined to become a seed would be fertilized by a sperm from a different plant and that thus the amount of crossbreeding would be increased. That was, at the time, highly desirable because the mixing of heredities increased prodigiously the number of variations from which natural selection (aided perhaps by some other less understood processes) could pick out for future use those which had survival value or possibly (as heretics suggest) even a value of a different sort. Flowers developed because they promoted this same "crossbreeding" even more successfully and one might reasonably expect that any further novelties nature had in store for us would continue to represent the general movement in this direction.

The fact is, on the contrary, that certain highly evolved modern plants began to unlearn their lesson. They got into the habit of permitting the male organs in one of their flowers to fertilize the ovary in the same flower. Heredities were no longer so well mixed, the tendency to vary was reduced, but the seeds were perfectly fertile and, so far as mere survival is concerned, those plants which had already evolved to a point where they were successful organisms flourished as a species perhaps even more abundantly than they did when cross-fertilization was the habit. Many of them still bear attractively colored flowers but their beauty no longer serves any utilitarian purpose. If they are visited by insects, the visits no longer have any biological consequences. And

if the orthodox evolutionists are right in believing that
nature never supports indefinitely any organ which serves no
purpose, then these flowers may disappear in time.

Now these functionless flowers and these plants which
have returned to a more primitive form of sexuality are remarkable enough — perhaps one might even say ominous
enough. But the dandelion that we have picked out for
our special attention is one of the very few plants that has
gone these others one better (or one worse), because it has,
as we said in the beginning, abandoned sex entirely. Its
ovaries are not fertilized by pollen from a stamen in the same
flower. They are not fertilized at all. No sexual process takes
place. Every seed and therefore every new generation is the
product of a virgin birth. For good or ill dandelions have
said goodbye to sex.

Offhand most of us would have been inclined to say that
sex is necessary to reproduction — at least among the
"higher" plants and animals. But as every biologist knows
and as the dandelion demonstrates, it isn't. Reproduction is
not what sex is "for." Its biological function is the mixing of
heredities, not reproduction. Indeed, we might say that
what it actually does is not permit but prevent "reproduction" — if by that you mean complete duplication. Without
the intervention of sex every plant or animal would be almost exactly like its parent and if there had never been any
such thing as sexuality evolution would have had so little
variation to work with that today we might all still be
protozoa or at least some sort of very simple animal.

Sexuality made the dandelion what it is. The abandon-
ment of sexuality will keep it almost precisely that. Ob-

viously, then, the modern dandelion is paying a biological penalty. If it lasts for another hundred million years it will probably change very little and "improve" not at all. Dandelions seem to have said to themselves, "We are good enough. There is no reason why we should want to change. We don't believe in 'progress' any more. We will stop right here. And that being the case, we don't have to bother with sex any more."

If the dandelion really does prefigure the plant of the future then, for plant life at least, nature is giving up not only sexuality but along with it one of the Grand Principles that have dominated evolution. What the vitalist calls "nature's passion for change and improvement" and a mechanist would describe as "a set of fortuitous circumstances favoring change" would then be actually only a passing phase and the next few hundred million years may be destined to be those during which the earth's flora is gradually dominated by plants which, in so far as they are capable of even very slow change, will change only in the direction of those simplifications which the high degree of efficiency they have already reached makes possible.

Whatever disadvantages may be attached to this ultraconservatism of the dandelion none of them affects survival. As anyone who has ever tried to keep dandelions out of a lawn knows, they are highly successful plants. They are not native to America but the Pilgrims probably had them almost as soon as they had meadows, because the dandelion has followed European man to every temperate region he has colonized. They are inheriting the earth. Few if any other plants are so obviously fit to survive. Or are more likely to survive longer. By our standards of value an orchid may be "higher" than a dandelion, but if it is really true that nature knows no value other than "survival value" the dandelion is among the highest of all plants and may safely be assumed to point toward the future. Perhaps as the millennia roll by the whole surviving flora will become more and more like dandelions, while other plants "less fit to survive" go to the wall before the efficiency of those which have discovered the dandelion's secret.

Possibly the very conservatism which the sexlessness of the dandelion entails will prevent it from dispensing very rapidly with the now useless yellow petals it still produces and which redeem it with a coarse kind of beauty. But some of its cousins among the composites have already succeeded in getting rid of anything that we recognize as a flower. In the dandelion, petals now constitute a conspicuous expenditure of energy on a feature that performs no useful function. If nature never permits such waste for long, then even the conservative dandelion may manage to abandon it.

Furthermore, if the road the dandelions have so successfully taken should turn out to be for all living things the road of the future, then life may be saying farewell to many different kinds of flowering. Blossoms may turn out to have been merely a passing phase in plant development. So, for that matter, may the other kinds of flowering turn out to have been merely a passing phase in the development of animals as well — even in the development of man himself. Natural selection, if it really is the only effective dynamic principle in the world, will not keep dandelions — or anything else — pretty just to please us.

Despite her former tendency to exploit all the possibilities sexuality presented, moral and aesthetic as well as utilitarian, it is obvious that Nature has recently been at least experimenting in an opposite direction and in some cases has tended to abandon sex as no longer useful in the pursuit of that mere survival which, so many biologists assume, is the only thing her mechanisms can permit her to value. Perhaps no animal exemplifies either the tendency or the success that sexlessness seems to bring so startlingly as the dandelion does. But the highest insects — which seem as likely as any other living creature to survive all others — have achieved great efficiency by denying sexuality to any except a few specialized individuals, and have thus deprived most individual members of the species of an opportunity to participate in any of those elaborations upon the sex instinct which proved so significant in the development of what we, if not nature, tend to think of as the "higher" manifestations of life. Those practical moralists who used to bid us learn industry from the ant ought now, to be consistent, advise us to observe also how much more efficiently a society can operate when romance, sexual pleasure, or even the desire for offspring of one's own do not intrude.

Volvox reminded us how quietly so tremendous a thing as sex could slip into the universe. The dandelion and the ant seem to warn us how quietly it might slip out again. Should it do so, then color and perfume would disappear from plants, love and all the phenomena associated with it from the animals. Moreover all consciousness and all intelligence might disappear with them, if intelligence and consciousness really do, as some insist, function only as responses to novelty and to alternate possibilities.

No premise of the most narrowly orthodox biology gives us any assurance that precisely this situation might not ultimately evolve. "Progress," so it proclaims, is inevitable. But biological progress cannot mean anything more than a closer and closer approach to perfection in those features of a living organism which have "survival value." If flowerless plants and automatic insects really do — as seems at least possible — have an edge on everything else in "the struggle for existence," then they should in time supplant all other organisms. They are the "fittest," and it is the fittest who must inevitably survive.

Ask anyone who professes this ultra-orthodoxy whether a dandelion or an ant is conspicuously "fit" for anything *except* survival and he will probably either look blank or, if he is ever so slightly tinged with philosophy, somewhat irritated. What else, he will demand, is there to be fit for? What does, what can, any organism want except to survive? Nature, he will add, is not sentimental and only a sentimentalist abandoned to meaningless subjectivity would ever talk about "beauty," "nobility," or anything else unless it has some demonstrable survival value.

As one leading American psychologist who prides himself on having got rid of all nonsense about "value judgments" and their necessity for human beings has put it: "The only value judgment which nature adopts for you [is] the factor of survival. . . . The one criterion which is thrust upon us is whether the group which observes a given practice will be here tomorrow." And if one adopts that thesis then the dandelion, which seems more likely than the orchid, and the ant, which seems more likely than the human being "to be

N

here tomorrow," are to just that extent the "higher," the "more fit, and the more "successful" organisms.

A civilization as bedeviled as ours by immediate threats to its existence is not likely to worry much about what might possibly happen during the course of the next few hundred million years. Any writer who wants — as so many writers apparently do want — to make his reader's flesh creep has more promising subjects ready at hand. My intention is not to regard the remote future with alarm, but to suggest some of the paradoxes the orthodox view of evolution involves, especially to note the impasse we reach when we try to base our own philosophy on what is assumed to be nature's.

According to the early Darwinians, progress would presumably continue to be as inevitable as it had always been. Given time enough amoebas just couldn't help turning into men, because every accidental variation in the direction of increased "fitness" was mechanically preserved and passed on by "natural selection." Given even more time man might possibly turn into something else. But in any event the movement was always onward and upward. And though this is still the orthodox doctrine, perversely ingenious philosophers began quite a long time ago to raise annoying questions.

Henry Adams, for example, introduced intellectuals to a certain ignoble creature called Sacculina, which happens to be one of the several hundred different organisms parasitic on the crab and affords an example even odder than the dandelion's of what the biologist is compelled to call "progress."

Remove an adult Sacculina from a crab's insides and it is

nothing but a mass of hungry cells absorbing from the crab's internal fluids the nourishment it needs. Presently, however, it will develop male and female sex cells, the male cells will fertilize the female, and from the resulting eggs will hatch a tiny free-swimming crustacean with an outside skeleton, legs, a heart, and even eyes — none of which were possessed by its degenerate parent. This relatively quite respectable infant will swim about looking for a crab and when it finds one will enter the crab's blood stream. After it has been here and there for a while it will finally settle down just under the crab's abdomen and gradually lose all the organs that formerly marked it as a rather highly organized animal.

"What a beautiful example," said the evolutionists, "of the law that organisms repeat or recapitulate during their development toward maturity the stages by which the species evolved." Human embryos have gill slits to demonstrate that the ancestors of the human being once breathed water. The larval Sacculina demonstrates that once, long before it developed parasitic habits, it was a normal crustacean rather like the "water fleas" that swarm in dirty ponds and are prized by aquarists as food for tropical fish. But the parasite is "higher" than the complete animal — unless you believe that there are values other than "survival value" which can be used as criteria.

Thus we go round and round in a dismal circle. Progress is inevitable because whatever happens is, by definition, progress. The fit survive because whatever survives is, by definition, fit. Modern animals are "higher" than more ancient animals because "higher" means "more recent." You may say that certain characteristics, endowments, and powers which have no demonstrable survival value seem to you "higher" than certain others which have. You may even say that certain organisms which have disappeared or seem likely to disappear seem to you more fit to survive than some which are inheriting the earth. But this, reply the orthodox, is mere prejudice and delusion. Nothing that is fit to survive can perish; nothing that does perish was really fit for the only thing which counts.

Even today there are those who profess themselves content to remain within this circle. Such a person was asked in the course of a public discussion whether or not he would regard with complacency an evolutionary development in the course of which man himself learned how to be, like the ant, so mechanically and unchangingly efficient that all his intelligence and even his consciousness would fade away. He replied that he did not anticipate such a development but that if it should occur he saw no reason for being disturbed. Such a creature would be all the surer of "being here tomorrow" and that was, after all, the only criterion which could be applied.

To break out of the dismal circle one or the other of two assumptions must be made. If you insist upon believing there is something for which Sacculina, for instance, is less "fit" than the ancestor from which it evolved; that other creatures are, in some meaningful sense, "higher," no matter how successfully Sacculina may survive — then you must believe also that there is some valid standard other than "survival value" by which "higher" and "lower" or "fit" and "unfit" may be measured. You may believe that during the course of evolution Nature herself has demonstrated that this other system of values is one she herself recognizes as valid. Or you may insist that to put some value above survival is something of which man alone among the creatures on this earth is capable. But believe one or the other you must.

Despite Sacculina and a number of other examples of what a human being must think of as devolution, the main movement which evolution itself describes has been in the general direction of organisms "higher" by human standards and more fit than earlier organisms were for various things other than mere survival. Recent animals are more intelligent, more intensely conscious, more capable of play, and joy, and love than the ancient ones were. And since it is impossible to demonstrate that they survive any more successfully or are any more likely "to be here tomorrow" because of these capacities then does not their emergence make out a very strong case for the assumption that nature values them — even if biologists sometimes do not?

If man himself is the product of an evolutionary process, then how could it have happened that he came to value things to which the process that created him is indifferent? Was he divinely set off from the rest of nature? Or did he, on the other hand, set himself off — not divinely but unnaturally and perversely? Is he merely something tolerated briefly by Nature before she gives the world wholly over to the dandelions, the ants, and the Sacculinas, from whom both the ability ever to change themselves further is slowly withering away along with the ability to love and to be beautiful as we understand those terms?

Perhaps. But there is another "perhaps" which seems at least equally probable. Perhaps the ant, the dandelion, and the Sacculina represent not the road of the future, but merely blind alleys. Perhaps they are only experiments ending in ignoble failure. Perhaps it is they who will be left behind to die out or even to survive shamefully, while intelligence, beauty, and everything which goes to make up the vivid consciousness of being alive goes on surviving and increasing — whether these characteristics have any "survival value" or not.

If natural selection, as the Darwinians understood it, is the only process affecting evolution, then it is hard to see how this could be. But if you permit, even as a minor influence, the intervention of "will" or "preference" on the part of the evolving organism itself, then there seems no reason why it could not.

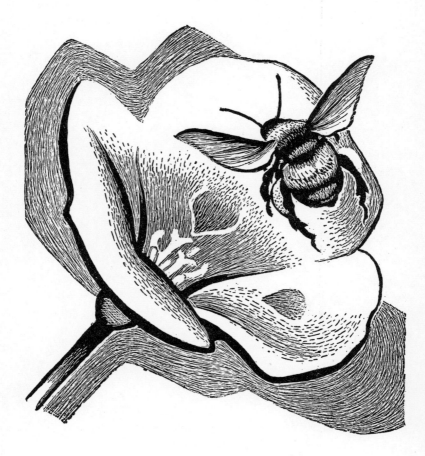

11. How Right Was Darwin?

To THE BANAL REMARK that "Life is strange" a wit once replied with the impudent query, "By comparison with what?"

Nothing remains to be said if the original remark was intended as a comment on some happy accident or unhappy contretemps of daily life. But if one is thinking instead of the whole phenomenon called "living" as it manifests itself in animate creation, then a meaningful retort is possible. One may say quite simply, "By comparison with everything that does not live."

Despite all the comparisons that have been made between organisms and machines, despite all the talk about "improbable chemical reactions," the difference between the animate and the inanimate, the discontinuity of the living and the nonliving, remains absolute.

Consider a "frost flower" on a window pane and the curling leaf of a fern. How similar at first glance they appear!

The forms or designs are startlingly similar. Both "grow" and seem to express some inner conviction or law. One might almost suppose that one had imitated the other. Yet they represent the two things in the whole universe which are the most different.

But how, after all, shall we define the difference? It is not, as we have just seen, merely a matter of "growth." Neither is it a matter of "being organized." The frost flower has form and shape. Its seeming organization is, if anything, the more "perfect." In its own dead, cold way it is almost as beautiful. And it is certainly far more ancient. Millions — actually billions — of years before the first humble blob of protoplasmic jelly was "alive" on our earth, frost flowers had been "growing" just as well and just as perfectly as they grow today. Quite possibly they will go on doing so for billions of years after life has ceased to exist on the planet. As long as there is moisture anywhere and cold to chill it, frost flowers will certainly grow. They cannot become extinct and they cannot change. They will always be what they have always been.

The fern, on the other hand, has not been here very long as geology measures time and it may not be here very much longer. It is true that something very much like a present-day fern was already growing when the coal beds were being laid down and long before any plants bore flowers, as well as long before most of even the "lower" animals with which we are familiar had come into existence. But for all that it is recent, changing, and temporary. By comparison with a frost flower it is tender and ephemeral.

Yet there is something about the fern leaf which stirs

us — or at least some of us — as the frost flower cannot. We can love it in a different way — not as pure Beauty, but with a love which is impossible without something like sympathy or fellow feeling. And we can sympathize because we know that the fern is like us, while the frost flower is not.

How shall we define that "something like us"? Is it merely that we are both perishable intruders into a dead environment over which we both, temporarily at least, are able to triumph? Or is it something still more than this? Is it because we can say — and mean something when we say it — that the fern "wants" to be left alone in its cranny by the rock? Certainly many people do have some such feeling and they are right to have it. In some sense all living things are allied in some sort of struggle against all that are not living.

Believing that everything about him was alive, primitive man attributed a psychic life to mountains and winds, to rivers and stones. No doubt the distinction that was slowly made between the living and the inanimate was tremendously important in defining his own mental world, because it tended to draw him emotionally closer to other living things while it marked him off from whatever did not live. But it is a curious aspect of modern intellectual development that modern thought has, on the contrary, tended to obliterate again the distinction, to interpret life in mechanistic terms until, by now, it might almost be said to have come to a conclusion exactly opposite the assumption of primitive man. If the latter thought that everything in the universe was alive, the mechanist believes that nothing is, and the significance of even the word "organism" as distinguished from the word "machine" tends to disappear. Moreover, and as the result of a somewhat similar development, the medieval man who saw "purpose" everywhere and, for the most part, purpose directed toward him and his needs, has given way to the mechanist who sees purpose nowhere and rejects the assumption of even the most generalized "intention" in nature almost as vehemently as he rejects a naïve, man-centered teleology.

Many biologists have moments when they acknowledge the ultimate mystery and wonder of life but often they are too irrevocably committed to mechanistic dogmas and too afraid of the sneers of their fellows not to hedge even when their own logic compels them to admit that the accepted premises are by no means wholly satisfactory.

Consider for example a brief quotation from the preface to the most recent synthesis of biological knowledge, the eighteen-volume *Traité de Zoologie,* which began to appear in 1952. The preface in question was written by the general editor, Pierre P. Grassé, a professor at the Sorbonne, and I risk a translation interrupted by a few bracketed questions.

Biology, despite the brilliance of its appearance, stammers in the presence of the essentials. We know neither all the properties of living matter, nor all of its astonishing possibilities.

It is certainly ruled by laws as strict as those of inert matter [How is that "certainly" justified?], but the diversity of its elements and the complexity of their architecture, fixed and moving at the same time, generate at its heart capacities whose originality never ceases to surprise us; such as assimilation and the power of regulattion. [Why not add the greater surprise of consciousness which certainly emerged sometime whether or not it extends far down the scale of organisms?]

Psychic processes and living matter are intimately, indissolubly connected. [Intimately certainly; indissolubly perhaps. But how can we be sure of the second?] The amoeba, apparently so simple, comports itself in a way which outlines and at times fills in the features of the behavior [Why only of the behavior?] of higher animals. All living matter which has assumed unitary form, the form that is of an individual, has its own psychic aspect. We must accept this evidence whatever may be its philosophical consequence.

But what are these "philosophical consequences"? Do they confirm or even make probable any theory which refuses to accept the radical discontinuity of living with inanimate bodies?

Even so cautious a statement as that made by Professor Grassé carries us a long way from the phrase "wound up like a watch" which was so confidently employed to describe living organisms by the first enthusiastic mechanists. But most biologists hesitate to take one step farther because, so they say, you cannot take it without falling into the arms of Lamarck and Vitalism, both of which are, they will add, thoroughly discredited.

Obviously, the speculations in which this book has indulged have sometimes seemed to beckon from across the forbidden line, and in all humility I should like to defend them as legitimate speculations.

That a mere layman has no right to make authoritative pronouncements is granted. But since the authorities do sometimes seek to convince him he ought to be permitted to confess that of certain things he has not been convinced without being visited with the wrath of an *odium theologicum.* In science no less than in religion honest doubt is worthy of respect and it is for honest doubts that I plead.

This being the case, it is well to be as specific as possible. Here, then, are things which as a layman I have been able to believe or not believe: (1) That the higher plants and animals "evolved" from the lower seems to me as certain as circumstantial evidence can make anything. It is, to use the legal phrase, beyond reasonable doubt. (2) That natural selection, working mechanically, has contributed very importantly to the process seems almost equally certain, because the circumstantial evidence is again very strong. (3) On the other hand, the statement that no factor except natural selection has ever influenced the course of that evolution seems to me a statement based wholly on negative evidence and one making the whole story so nearly incredible that negative evidence alone is not sufficient to rule out completely certain alternate possibilities.

A doubt that orthodox Darwinism really does embody the whole truth was expressed by Alfred Russel Wallace from the beginning; and similar doubts about the adequacy of the various neo-Darwinisms which have followed upon the original theory have been shared by respectable thinkers, especially outside the ranks of professional biology — notably by Samuel Butler and Bernard Shaw. Some scientists are inclined to dismiss such doubts as "mere literature," but even if that be accepted as a sufficient condemnation it is worth noting that doubts have been spreading even into scientific circles.

In her *Man on Earth* (1955) the Cambridge-trained archeologist Jacquetta Hawkes was moved to ask: "Is it possible today for any unprejudiced and intelligent person to believe in the orthodox view of the workings of evolution? Can such a person believe that we and all our fellow travellers in life from apes to algae have been shaped and colored and endowed with our highly distinctive habits by natural selection alone?" Even more recently Edmund W. Sinnott, botanist, geneticist, and dean of the Yale graduate school, has raised in *The Biology of the Spirit* the whole question of the seeming duality of matter and spirit, has asked whether or not one is "no more than an illusion" and expressed the final opinion that in protoplasm itself there exists a "principle of organization" that "brings order out of randomness, spirit out of matter, and personality out of neutral and impersonal stuff." Another academic biologist, G. M. McKinley, professor of zoology at the University of Pittsburgh, has also joined the heretics with *Evolution: The Ages and Tomorrow*, in which he dismisses as absurd the whole attempt to account for consciousness and intelligence in either man or the lower animals by natural selection alone. Obviously then, doubt can be respectable (if somewhat unusual) even within the circle of professional biology.

Bernard Shaw has often been ridiculed for saying in effect

that orthodox Darwinism simply cannot be true because it is too immoral and too dispiriting; because it teaches that never, since the beginning of time, has anything, from amoeba to man, been able to improve itself or to influence its fate. But at least his attitude calls attention to the fact that the question at issue, far from being of merely technical interest, has consequences very important for the society which answers it.

Both the conduct of modern man and his attitude toward the universe in which he lives have already been profoundly affected by his readiness to believe he is only a machine that created itself by purely mechanical means; that his convictions are the result of what happens to him rather than that what happens to him is in part the result of his convictions; that both "purpose" and "value" are, at most, insubstantial creations which have no counterparts anywhere outside himself. If you believe that, then the whole universe of which you are a part becomes a mere machine, not really alive in any sense usually associated with the term. Reject it, as Shaw does, and the universe becomes alive again.

If the universe really is alive, then one is free to believe also that living creatures were from the beginning endowed with some power to intervene in the evolutionary process; that it was not exclusively an external force which moved them upward toward greater complexity; that their own dim minds, dim wills, and dim preferences helped them along the way. Indeed, one may then even translate this modest statement of the role of something not blind and mechanical into the rhetorical form Shaw chooses in *Back to Methusaleh* when he makes the Serpent say that the two great instruments of change are Imagination (which enables us to picture what might be) and Will (which enables us to see to it that the possible shall become the actual).

To all this the orthodox are likely to reply simply: "Very pretty. Perhaps it would be nicer if something of the sort were true. But it just happens that it isn't and that modern science has conclusively demonstrated the fact." This brings us straight to a question of our own: "Has it? Upon what kind of evidence does the statement that natural selection *alone* is the cause of evolution really rest?"

That the case for natural selection as a factor — and probably a very important one — is almost unassailably strong I have already admitted. It does take place, and unaided it may possibly be capable of producing organisms better fitted for survival than their parents. But whether or not it alone is sufficient to account for *everything* that has happened is quite a different question, and Darwinians themselves have never been entirely comfortable with their own interpretations.

No one today would claim that the theory as Darwin first set it forth is adequate. Further studies have modified the original conception and discovered, or imagined, new ways in which it might be supposed to work better. Such is, of course, the usual history of scientific theories. But the repeated modifications of the original theory do testify also to a certain gnawing doubt whether the theory as it has stood at any one time was fully adequate. Time and again it has been tinkered with, modified, made more elaborate. Yet early Darwinians had accepted it as a dogma not to be

O

questioned before it had the support of later discoveries. They asserted its adequacy in a form no one today believes adequate. And if they believed it before it was credible, then can one be certain that it would not be generally believed today even if it were not, at last, unassailable?

Even now the tale as told by the most careful expositors is a very tall one and there are few who will not admit that it would be easier to swallow if only we could believe that it was not exclusively a story of the miracles which pure chance, complete blindness, and utter mindlessness are supposed to have accomplished. But it is usually added that facts are facts and that no intervention of will, purpose, or intention did occur — or at least did not until man or some near-man had emerged with the mind which had been mindlessly created.

But why and how justifiably can anyone be so sure? Because, so the orthodox will reply, mechanical selection can be discovered at work whereas nothing except mechanical selection ever has been so discovered. Yet this argument is purely negative and it overlooks an important fact, namely, that mechanical operations are relatively easy to observe, whereas those of the will or the spirit would be difficult if not impossible to detect by the methods commonly used.

Take for example one of the classical experiments which has been repeatedly described. You tether somewhere various individuals of the same species of insect or small mammal which vary markedly among themselves in color. Presently it is found that those which blend with the background are picked off less promptly by birds or other predators. Obviously, then, protective coloration has survival value.

The longer an individual survives the better his chance to reproduce and to pass on to another generation his advantageous color. Hence, so it is concluded, natural selection has produced protective coloration as well as all the much more striking examples of disguise and mimicry nature provides.

So far so good. That the protectively colored insects and the sand-colored mice in desert regions were produced by this process is credible enough. The creation by the same process of a mind that devises such experiments and ponders the conclusions to be drawn from them is not quite so easy to believe in! But that is not the principal point. The principal point is that the methods employed in the classic experiments are such that they are necessarily incapable of detecting anything except what they do detect.

The proof that in these cases natural selection actually does select is purely statistical. But statistical proof never can eliminate the possibility that factors other than those which are the subject of the experiment exist. Suppose that some individual insects or mice were in some way aware of the advantage selective coloration gave them. Suppose they could give themselves a slight push in that direction. The fact would not be detected in the course of the experiment. It would increase somewhat the survival rate of those favorably colored. But since you do not know what that rate would be on the basis of the merely mechanical assumption, the contribution of the nonmechanical factor would not be detected.

Grant that in this particular case the known factors do seem as though they might be sufficient to explain the facts and that there is no reason for suspecting the existence of any others. Even so one should nevertheless admit in all honesty that their nonexistence is not conclusively demonstrated even though one might let it go at that. But when it comes to the whole story of evolution, the situation is very different. It is a much taller story and one is less ready to believe that natural selection alone is sufficient to account for it.

In this case, then, we have a far better reason for caution and some real excuse for the suspicion that some other factor has been at work. The mere negative argument against the assumption is no longer sufficient. There is no longer any excuse for denying positively, dogmatically, indignantly, and even irascibly that natural selection is the only possible explanation of everything which has happened to evolving life since the beginning of time. Doubt is here at least legitimate and respectable.

That such doubt has been expressed by a Samuel Butler and a Bernard Shaw then becomes, not proof that they are mere literary men from whom logical thinking is not to be expected, but proof that they take advantage of holes in the logic of their opponents which those opponents will not, or cannot, see. Recent philosophers have listened very respectfully to scientists. It is not certain that the latter should not return the compliment.

Unfortunately, however, all tendency to question in any way the adequacy of the purely mechanistic theory of evolution is now labeled either Lamarckian or vitalistic; and no less unfortunately both these words have become epithets to be hurled rather than terms indicating a position to be analyzed. To raise such reasonable questions as we have been raising is to be labeled Lamarckian and vitalist and to be so labeled is to be accused of holding two absurd ideas: (1) that life is something distinguishable from the matter which a living creature animates, and (2) that a giraffe, to take one example, got a very long neck because for a long time he had been thinking how convenient it would be to have one. Yet it is quite possible to believe in the reality of something besides natural selection as a factor in evolution without deserving either of the labels or taking any position fairly stated in terms so simple as those just used.

It is true that Shaw himself used the extravagantly comic example of the giraffe — whether he took it seriously or not. But there is a considerable difference between so simplified a fable and the suggestion that choice, consciousness, awareness, and will have sometimes been able to intervene in an evolutionary process which has, therefore, not been wholly the result of blind accident. Such a suggestion does not at all depend upon any assumption that acquired characteristics are inherited and does not, in fact, imply anything more than that the generally admitted ability of the human being to intervene may be assumed to have existed to some extent for a very long time — perhaps even for as long as any form of life has existed. And if to suggest that as a possibility is in itself sufficient to earn the label "Lamarckian," then perhaps a Lamarckian is not obviously an absurd thing to be.

The term vitalist and the nonsense it is supposed to suggest is not so easily disposed of, because the doctrine it implies is elusive — so elusive in fact that one may easily begin to suspect a mere dispute over words. What it commonly implies is the notion that Life is something different from and separable from any body it happens to inhabit. Such a notion leads naturally to the assumption that life may exist independent of a body and therefore suggests even the possibility of both pre-existence and individual immortality. Thus to say that a thing is alive seems to come very close to being the same thing as to say "It has a soul." And it is this which so outrages the orthodox scientific mind.

But in acrimonious controversy the scornful epithet "vitalist" is too loosely applied. One is not necessarily a vitalist just because he feels, as I do, that those who dismiss the life processes as "mere chemistry" and speak glibly of "the improbable chemical reaction which occurred only once and has propagated itself as life ever since" is to go far beyond any demonstrable facts. Neither does one necessarily become anything of the sort by going as much farther as I have gone in insisting that, for the present at least, less violence is done to known facts by acknowledging rather than by refusing to acknowledge that the simplest living creatures already exhibit capacities radically different from any which a mere chemical compound ever hints at. To say this is not to say that life has ever existed separate from a body or that it can so exist. I doubt that it ever has. But I do wish to preserve in discourse a distinction which, so far as we know, is always present in the phenomena we discuss.

A physicist such as Professor Gamow can permit himself to say that the "basic manifestations of life like growth,

motion, reproduction, and even thinking depend entirely on
the complexity of the molecular structures forming living
organisms, and can be accounted for, at least in principle,
by the same basic laws of physics which determine ordinary
inorganic processes." On the other hand, Dr. Robert Oppenheimer has recently given it as his opinion that "it seems
rather unlikely that we shall ever be able to describe in
physicochemical terms the physiological phenomena which
accompany conscious thought or sentiment or will." And
he has added: "Should an understanding of the physical correlate of elements of consciousness indeed be available, it
will not be itself the appropriate description for the thinking man himself, for the clarification of his thoughts, the
resolution of his will, or the delight of his eye and mind
at works of beauty."

This does not make Dr. Oppenheimer any more of a vitalist than Professor Gamow is. But it does keep him closer
to known facts. When he speaks of the "physical correlate"
of life rather than of its "physical equivalent" or "cause" he
makes the distinction I have tried to maintain. And when
he adds, in effect, that to describe the physical correlates is
not to give an appropriate description of psychic events in
the human mind he is saying as much as I have meant to
imply — at least if Dr. Oppenheimer would be willing, as
he might well be, to extend the statement to include psychic
events in all living creatures, human or subhuman.

Readers may have noticed that in previous chapters I
sometimes seem to vacillate between two points of view,
speaking now as though life were a mysterious indescribable

potentiality in matter, now coming close enough to vitalism to suggest that it is not material at all. But this seeming inconsistency arises out of a verbal difficulty created by the fact that our ordinary vocabulary makes distinctions which science itself is now exposing as probably not real.

Both this inconsistency and, indeed, the difference between "mechanism" and a good deal of what the mechanists brand as "vitalism" disappear in the light of recent developments in the science of physics. That science has found itself more and more concerned with the hitherto unsuspected attributes of matter which reveal it to be not as completely subject to mechanically operating laws as was formerly supposed. It has found itself at the same time compelled to recognize that since no necessarily enduring distinction exists between matter and energy, the distinction between "material" and "nonmaterial" has ceased to exist. Hence, whether you say that life is something merely mechanical and merely material or whether you say that it seems to represent a potentiality of matter previously as unsuspected as those potentialities with which recent physics has been dealing, makes very little difference.

I find it impossible to believe that Newton's billiard-ball atoms, whose every movement is predictable according to immutable mechanical laws, could be so organized as to produce consciousness and thought. But the atoms of modern physics have become both totally unlike Newton's and confessedly indescribable. What they are capable of no one can imagine. Thought itself may be simply one of the attributes of matter which becomes manifest when atoms are organized in some particular way. And if that is true, then the life which at some time appeared on earth in the form of a sub-amoeba may have been very different from that merely "material" thing which old-fashioned mechanists imagined it. Perhaps neither matter nor the universe composed of it were ever "dead" in the sense they were once supposed to be.

Old-fashioned materialists used to ridicule the idea that a mere collection of material particles could ever become capable of exhibiting such "spiritual" (and therefore illusory) capacities as are suggested by the words "choice," "will," and "moral preference." But organisms certainly are capable of consciousness and that is a hardly less improbable manifestation of mere matter than free will is. There is no logical way in which a materialist can escape vitalism except by admitting that the very characteristics which vitalists attribute to the thing they call life are actually potential in matter itself. And at that point the dispute between vitalists and materialists becomes a mere matter of words. Either life exists as something distinguishable from matter or matter itself is or can become "alive."

For more than a century it has been known that every chemical element found in any living organism is found also in inorganic matter. Recent proofs that the so-called elements are not really elementary but capable of turning themselves into some other so-called "element" and that all atoms are made by combining in different ways the same fundamental stuff merely means that the living and the dead are composed of the same *one* thing instead of the same series of different "elements."

Now, as before, the question remains a question of the

differing "organizations" which make the difference between iron and gold on the one hand and the difference between what is alive and what is not on the other. And the most momentous aspect of the question is this: Is that organization of matter which enables living things to live merely *more complex* than the organization of inanimate matter, or is it organization on some radically different plan? It is to this form that the old question of whether the organic and the inorganic are identical or different now reduces itself.

Most biochemists would, I believe, answer "merely more complex" — with varying degrees of assurance. But they do not know how, why, or at just what point this "greater complexity" endows the matter so organized with the characteristics which still seem to set off the living from the nonliving. And since that is true it seems to me that it becomes quite legitimate to speculate upon the suggestion that the two kinds of "organization" may actually be on some radically different plan. Moreover, there is one very important aspect of modern physics which seems to hint at what the consequences, if not the nature, of the difference between the two kinds of organization might be.

The great paradox at the heart of recent developments in scientific theory could be put in this simplest possible form: "Large aggregates of matter seem to obey fixed laws; very small bits of matter do not. Newton was right about all gross mechanical phenomena. Apples do fall and planetary bodies do obey his laws of motion. But he was wrong in his theoretical assumption that every particle of matter in the universe behaves in the same predictable way."

You cannot, say the atomic physicists, always predict what

an individual atom is going to do. The apparent predictability of apples and planets is actually based upon statistical probability, not upon immutable law, so that it is analogous to that predictability of the mass behavior of human beings as the result of which you can safely predict that fewer people will walk in a city park on a rainy day than on a sunny one — though you cannot predict that any individual man will or will not be there. In that sense freedom is an attribute of individual men, while necessity governs the behavior of the group.

Inevitably the revelation of what has been picturesquely called "free will among the atoms" had led to the suggestion that man also may be free. Dr. Oppenheimer — perhaps as a corollary to his reluctance to discuss psychic phenomena in physical terms — has been at some pains to express his doubt that the unpredictability of atoms really is relevant to the question whether or not human behavior is determined. Yet the fact that these individual atoms do seem to share with the human being a characteristic which is no longer evident when the atoms are organized in the form of inanimate matter suggests the possibility that such organization renders ineffectual a property which organization into living matter does not.

What, then, I am suggesting as a speculative possibility is simply this: The organization of atoms into nonliving material does differ radically from the kind of organization which results in living matter; and the consequence of the difference is that whereas the one kind of organization is such that the unpredictability of one individual atom cancels out the unpredictability of another, the other kind of organization results in a pooling or cumulation of the freedom inherent in the individual particle, so that though a baseball is less free than the atoms which compose it, a man is even more free than any one of the atoms of which he is composed.

What these two kinds of organization are like, how they differ, and why they produce these contrary results are questions upon which I am not able even to speculate. But that does not obscure the fact that some such description as I have given of the consequences which follow upon the one and the other does suggest a way in which the assumption that all matter is fundamentally the same may be reconciled with the fact that living and nonliving aggregates of that matter seem radically unlike. In some sense it reconciles the mechanist who insists that biology is merely a branch of physics and chemistry with those whom he accuses of vitalism because they insist that the life processes seem to involve something not involved in what are ordinarily called chemistry and physics.

Two very recent developments, the one theoretical and the other experimental, are relevant.

The distinguished chemist Irving Langmuir has developed for chemistry itself the concept of "convergent" and "divergent" phenomena, by which he means to distinguish those chemical phenomena in connection with which statistical methods increase predictability and those in which they do not. To acknowledge the existence of the second class of phenomena seems to reinforce the validity of the speculation offered above concerning the possibility that "animate" and "inanimate" may represent two different methods of organization. The psychologist Ivan D. London has approached more closely still the same suggestion in an article called "Quantum Biology and Psychology," in the *Journal of General Psychology* for 1952, in which he applies Langmuir's distinction to psychic and biophysical phenomena.

The other recent development is the announcement that a filterable virus has been resolved into its chemical constituents and then resynthesized into a virus again. The meaning of this experiment is obscure, but should it be confirmed it would certainly be taken as further evidence against vitalistic theories as they are commonly stated. On the other hand the experiment is essentially irrelevant so far as concerns the monistic concept, which proposes not that life be regarded as something separable from matter, but that life is something potential in matter just as, so we now know, matter is not something essentially different from energy but simply something which may become energy.

To say "I am a materialist" is meaningful only if the term "matter" can be defined in such a way as to distinguish it from something else. Until a few years ago such a definition seemed easy. "Matter is that which occupies space and has weight." This definition distinguished it as clearly from "energy," which occupies no space, as it does from such vaguer terms as "life" or "spirit" or "idea," which, having neither the characteristics of matter nor, demonstrably, the characteristics of energy, were sometimes assumed either to be an obscure form of energy or not to exist at all. To say "I am a materialist" meant, "I believe that the only fundamental reality is that which occupies space and which has weight."

But the matter that disintegrated privately over the

American desert and then publicly over Japan ceased in those instants to occupy space. At those instants, therefore, the meaning of the term "materialist" disappeared as completely as the disintegrated atoms themselves. Between the man who says "I am a materialist because everything is ultimately material" and the man who says "I am not a materialist because nothing is ultimately material" no definable difference any longer exists.

On the basis of this seemingly demonstrated fact, many a scientific treatise will have to be revised if the now meaningless statements are to be removed from them. Almost at random I open a recent book on one of the biological sciences. The authority of a distinguished scientist is quoted to support the contention that the first appearance of life on earth may be accounted for "without the intervention of the non-material." In the light of the most recent knowledge does this statement mean anything at all?

Such speculations as these at least remind us of the astonishing fact that after this earth had existed for billions of years there finally appeared upon it a creature capable of abstract thought, interested in such philosophical questions as science raises, and bold enough to be able, or at least fancy that he is able, to reach back through billions of years to form what he believes to be more or less correct opinions concerning what happened so long ago.

If it really is true that he is merely the inevitable culmination of an improbable chemical reaction which happened to take place once and once only and involved "merely material" atoms, then the fact that he has been able to formulate the idea of "an improbable chemical reaction" and to trace himself back to it is remarkable indeed. That chemicals which are "merely material" should come to understand their own nature is a staggering supposition. Is it also a preposterous one?

Without attempting to answer that last question one thing more may be said. If it should turn out that man has not *understood* but *misunderstood* his own evolution then that is still a fact almost as staggering as understanding it would be. But it is also a staggering irony.

If, to go one step farther, the misunderstanding should lead him to deny, disregard, and allow to atrophy through disuse the very characteristics and powers which most distinguish him — if, in a word, he should thus help himself back down the road he once came up — that would be more than a staggering fact and more than a staggering irony. It would be of all calamities one of the greatest that could befall him — greater perhaps than any except that possibility of falling into the hands of an angry god he once so much feared.

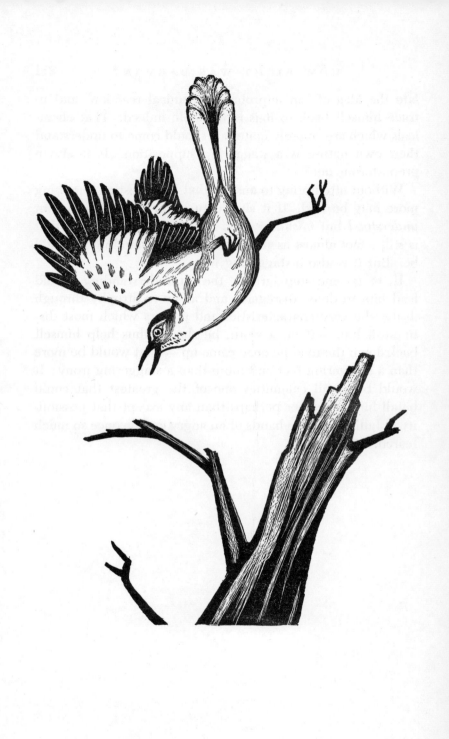

Epilogue

THE FIRST SENTENCES of this book were written nearly two years ago. Outside my window on that spring morning, as on this, a bird sang. Outside a million windows, a million birds had sung as morning swept around the globe. Few men and few women were so glad that a new day had dawned as these birds seem to be.

Because my window looks out on a southern landscape, my bird is a cardinal, with feathers as bright as his half-whistled song. Farther north in the United States he would be a robin, more likely than not — less colorful and somewhat less melodious but seemingly no less pleased with the world and his place in it. Like us, robins have their problems but they seem better able to take them in their stride. We are likely to awake with an "Oh, dear!" on our lips; they with a "What fun!" in their beaks. Mr. Sandburg's peddler was remarkable because he seemed so terribly glad

P

to be selling fish. Most robins seem terribly glad to be eating worms.

For some time I had been thinking that I wanted to write a book about the characteristics and activities of living things. During the week or two just before, I had been wondering with what activity or characteristic I should begin. Reproduction, growing up, and getting a living are all, so I said to myself, fundamental activities. Combativeness in the face of rivals, solicitude for the young, courage when danger must be met, patience when hardships must be endured, are all typical characteristics. But my cardinal proposed a different solution. Is any characteristic more striking than the joy of life itself?

No starting place is less usual or would have seemed less suitable to many biologists. Some would certainly prefer to begin with origins — with the simplest creatures now living or with the theoretically even simpler ones from which they evolved. Others might choose an abstraction, but the abstraction would probably be "the struggle for existence" or "the survival of the fittest." Pressed to name the most fundamental characteristic of life they would probably reply; "The irritability of protoplasm."

With them on their own ground I certainly had no right to quarrel. The cardinal and the robin do have to engage in a struggle for existence. The protoplasm in the cells of their bodies is, like that in mine, "irritable." But when I hear the word "robin" — especially when I hear a particular robin singing on a bough — I do not think: "Irritable protoplasm so organized as to succeed in the struggle for existence." I think that no more than when I hear my own name I think:

"Member of the American middle class, subdivision intellectual, caught in an economy where he is not very comfortable
and developing opinions which are the produce of his social
situation." An equally significant sort of fact about both
men and birds is that individuals are more or less happy,
terribly glad or terribly sorry to be doing what they are
doing, and capable of making more or less interesting comments on their situation.

With this fact science can hardly concern itself. Such
facts are not measurable or susceptible of objective demonstration. But to men and to robins alike they are nevertheless
very important and very real. If this were not so I do not
think I should ever have taken much interest in either human
or natural history. And if I consented in the end to begin
this book more conventionally, it was with some misgiving.

Men have surrendered a good deal of their capacity for
spontaneous happiness, and there may be compensations.
In any event our situation is one for which there is probably
no radical remedy. Yet even for us happiness is still important and it is, or at least once was, a fundamental characteristic of life. Nothing the lesser creatures can teach us is more
worth learning than the lesson of gladness.

Of this lesson the robin is an especially effective teacher,
for the same reason that certain men and women are. He
has, I mean, the gift of language. Even the happiest human
poets may be no happier than some of their less articulate
fellows but they are better equipped to communicate. Perhaps your robin and my cardinal are no more terribly glad
that a new day has dawned than the field mouse was when

the sun sank and the moon rose. But they put their gladness into sounds which are almost words. And unlike many other animals who make sounds, they speak our language. Their song is one we might, if we could, sing in some unreasonably happy moment. And it is this fact that makes us aware of them in a special way.

Perhaps it is no more than an accident. In no other respect are birds so much more like us than any other animals are. But language is one of the strongest of bonds, and long before the dawn of history men took an especially sympathetic interest in birds because they could understand what the birds were saying, and because it appeared, as in the case of so many beasts it did not, that they had a seemly and an eloquent way of saying it.

Sometimes we resent the fact that in human society so much honor is paid to those who are unusually articulate; that we pretend at least to honor the poet more than any other citizen; that even the mere gift of gab can open the way to fame, wealth, and power. About birds we might feel in much the same way. They sang for one another millions of years before any man overheard them, but they have profited perhaps unfairly from the fact that their song does communicate to ears so different from their own.

Even today they are less persecuted than any other small creature. Fewer people who see a bird say of it — as they tend to say of any other small animal they happen upon — "Here is a little creature who is alive and wants to live. Therefore let us kill it at once." And it is a significant indication of how much the vocal powers of the bird are responsible for the special regard in which they are held that laws protect "song birds" even though many so protected are only unmelodious relatives of those who actually sing.

Certain penalties, also, the birds have sometimes paid for making sounds so pleasing to human beings. Just because these sounds are inherently agreeable they can tickle the ear of men so self-absorbed that their imagination never carries them to the glad or sad singer himself and they regard him as no more than a mechanical music box which it is necessary to confine in a cage. At times it has even been the custom to put out the songster's eyes in the belief, true or false, that blindness improves his song. This can only mean that, in some limited way, beauty may be enjoyed without sympathy for or even curiosity about the living thing which produces it. Similarly, oriental potentates confine beautiful women in harems so that they may be conveniently "loved." Similarly, Ivan the Terrible, so it is said, blinded his architect to make certain that he could never build a basilica for any rival. Similarly, for that matter, ladies used to admire so much the color and texture of birds that they put them on their hats.

But whatever delights such aesthetes as these may know there are some of us who do not envy them. To us, hearing the song without communion with the singer is no better than listening to sounding brass and tinkling cymbals. When we listen it is less because our ears are tickled than because we rejoice to know that there is rejoicing in the world around us. We want to divine what the bird is saying to himself and to his fellows; to feel that some emotion, if not some thought, is communicated. We assume that in some general way bird language is translatable.

The language of music, even of the music addressed by men to their fellow men, is notoriously easy to misinterpret — especially when we try to hear in it not only the emotion itself but the specific occasion of that emotion and when we attempt, for instance, to distinguish between the joy of falling in love and the joy of spending a day in the country. That way lies the embarrassment of Mendelssohn's friend who congratulated him on having got the very spirit of the Highlands into his Scotch Symphony only to learn that what he had been listening to was the "Italian" instead.

Mistranslating the song of birds is at least as easy. Because most of them sing best at mating time we naturally assume that they are love poets almost exclusively. And Darwin, with his stress on "sexual selection," encouraged the interpretation. No doubt most present-day ornithologists are partly right when they insist that the cardinal or the robin we have been listening to is not singing primarily to charm his mate or to tell the world that love is sweet in springtime. In fact, so they say, what he really means is: "All other robins or cardinals take notice. I am already here. I have staked out a claim to a certain robin- or cardinal-sized territory large enough to supply food for me and my family. Trespassers will be prosecuted." But partly right is not wholly right and the joyful exuberance of the song, whatever its specific message, is as unmistakable as the fact that both the Scotch and the Italian Symphonies are cheerful pieces.

Yesterday, outside this same southern window beyond which the cardinal had been singing, I watched the antics of a mockingbird, who had been at these antics for many days. Perched at the top of a high tree he sang most of the day and through a large part of the night. Besides singing he leaped every few minutes a foot or two in the air, flapping his wings wildly and then settling back upon the same topmost twig. Probably even the most unromantic ornithologist would not deny that besides proclaiming his possession of a territory my bird was also trying to attract the attention of a female bird. Some of them might even adopt my fancy that, as the days have passed, there have been signs of a growing desperation and that the mocker seems to be saying to himself: "For goodness' sake, where are all the women? There are supposed to be enough to go around and I am certainly a pretty good specimen as mockingbirds go. I can sing long and loud. And I can jump as vigorously as anybody."

The most important thing is not the question whether the mocker's song is saying "I love you" or "This is my home and my land." The most important question is simply whether or not he is, as he sounds, confident and happy. And I am sure that he is. When a man tries to charm a woman by his conversation or when he describes the countryside in which he has settled down to live we assume that some emotions accompany his words. And whatever the bird is saying it fills his universe with joy.

Those creatures who cannot sing, or who do not speak our musical language when they do, communicate their joy in less direct ways. And the most eloquent of these ways is play. In some respects it is the most convincing of all the evidences of animal happiness because it demonstrates an excess of energy over and above what is required for the business of keeping alive. Those who study animals only in cages and laboratories know little about it. In prisons one must not expect to find much joy, human or animal. But the notes of field naturalists are full of accounts of the moonlight revels of rabbits and hares, of otters sliding down their chute-the-chutes into the water, of the gambols of the vixen with her young.

Only a few days ago two ground squirrels, so small that they must have just left their subterranean nests, came face to face outside my window. They touched noses, leaped each a foot into the air, and then scampered away in opposite directions "as playful as kittens." It makes no difference if you say that play is only "a preparation for the serious business of life." So presumably it is in the case of human children. But it is joy, not a realization of the necessity for exercise, which inspires the antics. Does anyone seriously doubt that gamboling children are having fun or that those emotions which accompany their play and constitute its meaning in their own immediate experience have nothing to do with joy?

In the room with me as I write, but confined to a roomy glass case, is one of those appealing little desert animals called a Kangaroo rat. I do not intend to keep him indefinitely because I do not like to keep "pets" who are not obviously as glad to stay with me as I am to stay with them. Nevertheless the Kangaroo rat is a solitary animal who, I like to think, is not lonesome in captivity or very much distressed by it. He spends a good deal of time pushing the sand about to make piles near his sleeping box, in filling his cheek pockets with the abundant food I supply, and in prac-

ticing that complete abstention from drinking which is his
chief claim to fame.

In his cage I put a little exercise wheel like that which
accompanies the old-fashioned squirrel cage. It took him
some two weeks to learn what it was good for. The fact that
he now races in it fast and expertly seems to me a considerable tribute to his intelligence, because the whole contrap-
tion is unlike anything for which his inherited reflexes could
have prepared him or his previous experience taught him
anything. But the real question in my mind was this: Does
he enjoy running in a wheel or does he use it *faute de mieux?*
Is this a kind of game or is it merely a poor substitute for
the exercise he needs and would have got in freedom?

Of course I don't know. But an anecdote told me by the
naturalist-photographer Lewis Wayne Walker after I had
begun to wonder about the matter suggests an answer. En-
tering his barn one night he was startled to see that an
unused exercise wheel stored there was revolving of its own
accord. He hid to watch, and presently a brown rat came
out, climbed into the wheel, had a fine run, and then went
away again. This rat could run as far and in whatever direction he wanted to run. He was not suffering from any en-
forced lack of exercise. But like a child to whom the family
automobile is no novelty but who nevertheless wants to ride
round and round on a carrousel, the rat was pleased for a
change to run fast without getting anywhere. It was fun and
that is all there is to it.

Perhaps man, beast, or bird does not find this kind of fun
so important or so satisfying as the sort which consists in
finding joy in an activity which has another purpose also — as

when a man and woman set up housekeeping or a robin sings to the world his happy awareness that he is in possession of the territory his wife and children have need of. Nevertheless, the ability to do something for fun, for nothing but fun, is a strong indication that this other kind of joy is also within the capacity of even a brown rat.

When I say all this I am not forgetting that many biologists would deny it almost *in toto*. Dogmatic "behaviorism" now has few adherents among those psychologists who are concerned chiefly with human beings, but it still dominates the thinking of many students of our fellow creatures. They may not go quite so far as Descartes and his disciples once did when they insisted that all animals other than man are mere machines incapable of even pleasure or pain because they are completely without consciousness. But they do cling to the contention that it is "not necessary" to assume any conscious concomitants of animal behavior and that since it is not "necessary" they will reject it.

Not many years ago one such naturalist was careful to explain to a popular audience that a bird singing on the bough must not be compared to a man singing in the bathtub. The man, he said, sings because he is happy; the song of the bird has "absolutely nothing" to do with "the joy of life."

Now neither he nor I can remember ever having been a bird. For that matter, however, neither he nor I has ever been anyone except our individual selves when we sang in a bathtub. But the assumption that the bird is joyous is very little less reasonable than the assumption that a neighbor who engages in melodious ablutions feels happy. In both cases we accept an analogy. We have no other evidence that human beings not ourselves are conscious at all. But common sense has always accepted the analogy of the bird and the analogy of the man as sufficiently persuasive. The man, we think, must be feeling something very closely similar to what we ourselves feel when we behave that way; the bird who sings and the animal who plays are probably feeling something at least remotely similar.

At least it is most certainly true that if the song of the robin does not express some sort of robin's joy analogous to our own then nature has no human meaning and we can study it only as we might study physics — merely because we are curious about how the machine works and may possibly learn things which will increase the efficiency of our getting and spending. Unless there is some emotion outside our own in which we can participate or from which we may draw comfort and joy then there is no universe beyond our own to which we can in any sense belong.

The ornithologist who has convinced himself that bird song "has nothing to do" with joy has not taken anything away from the robin. Ornithology notwithstanding, the robin continues to pour forth his heart in profuse strains of unpremeditated art. But such an ornithologist has taken a good deal away from himself and from those who feel constrained to believe him. They have forced themselves to live in a world that has come to seem, not joyful, but joyless. Robins and cardinals know better.

The gift for happiness is not always in proportion to intelligence as we understand and measure it. Birds are not as "smart" as dogs and monkeys are. But it is difficult to believe that even in liberty a monkey is as joyous as a bird or that he has the bird's special gift for gladness. Professor N. J. Berrill has put it thus: "To be a bird is to be alive more intensely than any other living creature, man included. Birds have hotter blood, brighter colors, stronger emotions . . . They are not very intelligent . . . [but] they live in a world that is always the present, mostly full of joy." More specifically Julian Huxley, surely no mere irresponsible sentimentalist, wrote thus after watching on Avery Island in Louisiana the love play of herons, who with loud cries of ecstasy twine their necks into a lovers' knot: "Of this I can only say that it seemed to bring such a pitch of emotion that I could have wished to be a heron that I might experience it."

The question whether a monkey can ever be equally happy is one upon which it would be better not to speculate. Inevitably it would lead sooner or later to another: "Can a man be as happy as a monkey?" And that had better not be asked. Perhaps some capacity for joy has been, must be, and should be, sacrificed to other capacities. Perhaps the happiness and solace which some of us find in an awareness of nature and in love for her manifestations derives in part from our imaginative participation in forms of existence from which the sacrifice of some of the capacity for joy in the interest of a capacity to think and a capacity to feel for others has not been very insistently demanded.

As for those who have never found for themselves either joy or solace in the teeming busy life which still animates those portions of the earth man has not entirely pre-empted for his own use, they might best be advised to begin by looking not for the joy they can *get* but for the joy *that is there*. And perhaps when they have become aware of joy in other creatures they will *get* by sharing it.

For me, at least, a bird is spokesman for more than merely himself and his kind. Just as individual men are accepted as spokesmen for their race and nation — permitted to sing its songs of triumph and of love — so the bird may be allowed to speak for other creatures. When I hear my cardinal I am reminded not only of all birds but also of all the furry and the scaled as well. Many of them are completely voiceless or so nearly so that they can only chirp or squeak. Others are so timid that I see their games rarely and by accident if at all. But the cardinal reminds me that many of these others might discourse of a joy perhaps almost as great if only they had the cardinal's gift of expression.

Nothing of what I have said need involve any sentimental unawareness of what is — to us at least — the tragedy and the cruelty in that same world my cardinal looks at and finds good. On yesterday's walk I came across the disemboweled carcass of a fawn, whose belly a coyote had ripped open. I needed no such reminder that even within the little stretch of earth visible from my window there are among the smaller creatures many such tragedies which occur daily and, especially, nightly. I see the hawk circling to kill and the buzzard circling as it hunts for the remnants of another's kill. In this world I have been celebrating, the fawn and the coyote cannot, by the law which imposes the contingencies of the natural order upon them, lie down together. If they did not accept this fact they could not be as joyous as they are. If I did not to some extent reconcile myself to it I might go on from protest to protest, until at last I must abhor the robin's slaughter of the worm as much as I did the coyote's slaughter of the fawn and end by finding both the life of man and all life outside him predominantly horrible.

Certainly neither the bird's world nor the world of any other creature is all joy. But even the question whether or not they have reason to be joyous is irrelevant. The tremendous fact remains that joyous they are whether or not it seems to us that they should be. Before our eyes they act out their joyousness and demonstrate very conclusively how fundamental a characteristic of life this capacity for gladness must be.

Perhaps joy is not so old as pain. Perhaps physical pain and physical pleasure are the earliest forms of awareness. But if joy is not so old as either, it may very well be older than sorrow because for sorrow we need a stronger sense of the past and a stronger sense of the future than most animals probably have. Sorrow is the child of Memory and of Anticipation, neither of which it is likely that my cardinal knows much about. Sometimes it is said that Eternity must be more like Now than like anything else we can imagine. If this is so then perhaps birds live in a series of almost discontinuous eternities. And many of them seem to be eternities of Joy.

Many reasons have been given by those who believe it a mistake for men either to create for themselves a wholly artificial environment or to remain unaware of the natural environment in which they live. The out-of-doors is said to be healthful for the body and tranquilizing to the spirit. Nature's ways are described as one of the richest subjects for the exercise of intellectual curiosity; knowledge of them is called indispensable for survival. All these reasons are valid. But none of them seems to me so persuasive as the simple fact that the lives of creatures other than man remind us compellingly of the fact that joy is real and instinctive. We have learned much that the animals do not know and developed many capacities they do not have. But they know at least one thing which we seem progressively to be forgetting and they have one capacity which we seem to be allowing to atrophy. To them joy seems to be more important and more accessible than it is to us.

Pleasure, which we seek as a compensation for the joy we so seldom feel, is both worth less and harder to come by. It requires some positive occasion and adequate occasions become harder and harder to create. Pleasure sickens from what it feeds on, joy comes easier the more often one is joyous. We relapse into melancholy or discontent and boredom. We suffer one or the other if we find at the moment no occasion for a different emotion. But nature, so it seems, relapses in joy. Is any other art more worth learning?